EXPERIENCES, INNOVATIONS AND ISSUES IN AGRICULTURAL EXTENSION IN UGANDA: LESSONS AND PROSPECTS

Experiences, Innovations and Issues in Agricultural Extension in Uganda
Lessons and Prospects

Editor
Margaret Najjingo Mangheni

FOUNTAIN PUBLISHERS
Kampala

Fountain Publishers
P.O. Box 488
Kampala
E-mail:fountain@starcom.co.ug
Website:www.fountainpublishers.co.ug

Cover, Layout: Robert Asaph Sempagala-Mpagi

ISBN 978-9970-02-715-6

Cataloguing-in-Publication Data

Experiences, Innovations and Issues in Agricultural Extension in Uganda: Lessons and
Prospects. – Kampala: Fountain Publishers, 2007

__ p; __ cm.

Includes tables and plates

ISBN 978-9970-02-715-6

1. Agricultural Extension, Uganda I.Experiences II. Innovations III. Issues

Contents

SECTION III: TRAINING AND CAPACITY DEVELOPMENT FOR INNOVATION AND REFORM IN AGRICULTURAL EXTENSION 85

Contributors

The editor

Margaret Najjingo Mangheni (PhD) is a senior lecturer in the Department of Agricultural Extension/Education at Makerere University, Uganda. Her experience includes four years as head of department and 17 years of teaching agricultural extension at Makerere University, and service as a consultant to international and national organizations in design, monitoring and evaluation of agricultural development programs. She has published over 20 articles in refereed journals, edited books, and proceedings of international professional meetings. Among her publications are the following co-authored papers:

"Information sources and constraints of service providers under the national agricultural advisory services program in Uganda" in *Uganda Journal of Agricultural Sciences. Vol 9 issue (11) p.p 257-264;*

"The relative importance of various skills and attributes for entry-level managers as perceived by agribusiness firms in Uganda" in *Journal of Extension Systems. Vol.21 (1), p.p. 83-95;*

"Gendering university curricula: Experiences of gender mainstreaming in the curriculum of the Faculty of Agriculture, Makerere University, Uganda". *In Tanzarn, N. (Ed) (2005). Gender in agriculture and technology. Kampala, Uganda: Makerere University Department of Women and Gender Studies, and Fountain Publishers, p.p. 127-146;*

"Challenges and policy actions for increasing the effectiveness of professional women agricultural extension workers in Uganda". In Tanzarn, N. *(ed.) (2005). Gender in agriculture and technology.* Kampala, Uganda: Makerere University Department of Women and Gender Studies, and Fountain Publishers, *p.p. 97-114.*

Email: mnmangheni@agric.mak.ac.ug

Chapter authors

Paul Kibwika (PhD) is a senior lecturer in the Department of Agricultural Extension/Education, Makerere University, Uganda, with over 10 years experience teaching at university level. He also has experience in facilitating processes for transformation of individuals and organization and has researched on learning for change; rural and agricultural innovations; local organizations development and knowledge management. One of his publications is a book entitled *Learning to make change: developing innovation competence for recreating the African university for the 21st century*, The Netherlands: Wageningen University.

Email: pkibwika@agric.mak.ac.ug

Monica Karuhanga-Beraho is a lecturer in the Department of Agricultural Extension/Education, Makerere University. She holds a Bachelor of Veterinary Medicine from Makerere University, and a Masters of Agronomy and Farming Systems (majoring in extension and agribusiness) from Adelaide University, South Australia. She has seven years experience teaching at university level and six years of community level agricultural extension work with Mukono District local government in Uganda. Some of her recent publications include: "Gendered impacts of HIV/AIDS and implications for food security" in *East African Journal of Rural Development, Vol. 21, p.p 137-17.*

"The women's movement in Uganda and women in agriculture". In Tripp, A.M and Kwesiga J.C. (eds.) (2002). *The women's movement in Uganda: history, challenges, and prospects.* Kampala: Fountatin Publishers. P.p. 90-105.

Email: monica@agric.mak.ac.ug

Emmanuel Beraho has a Bachelor of Veterinary Medicine from Makerere University and an M.Sc. in Agricultural Extension from Adelaide University, South Australia. He has worked with Uganda's National Environment Management Authority as a District Support Officer for the last 10 years, charged with strengthening environmental management capacity at local government level. He developed and piloted a District Environment Action Planning Framework for integrating environmental issues in district action planning in Uganda.

Jeff Mutimba (PhD), a Zimbabwean national, works for Winrock International as regional program coordinator, East and Southern Africa, assisting universities in the region to develop and run demand-responsive degree programs for mid-career extension professionals. In the past 11 years, he has assisted in developing and establishing extension programs at Haramaya and Hawassa Universities in Ethiopia, Bunda College of Agriculture in Malawi, Makerere University in Uganda and Sokoine University of Agriculture in Tanzania. He worked for the Ministry of Agriculture in Zimbabwe as an agricultural officer for fifteen years and retired at the level of under-secretary in 1989 to take up a position with CIATs as Regional Training Officer based in Ethiopia for three years. He then taught Agricultural Extension at the University of Zimbabwe for four years. With two colleagues, he has edited and published a book entitled *Intervention in smallholder agriculture: Implications for extension in Zimbabwe* to which he contributed four chapters. He has also published over 40 papers in journals, proceedings of international conferences, and other fora. He has served as president of the Southern and Eastern African Association of Farming Systems Research and Extension (SEAFSRE) and is the current secretary general of the African Agricultural Advisory Services (AFAAS) network.

Email: jeffmutimba@africa-online.net

Colette Harris (PhD) is a British national currently a research fellow in the Participation, Power and Social Change team at the Institute of Development Studies (IDS), University of Sussex, UK. In this capacity she teaches in postgraduate programs and carries out research, mainly on issues relating to conflict (from domestic through community violence to armed conflict) and family relations, all through a gender lens. She also works on public and reproductive health issues. In her previous position as program director for international women's issues at Virginia Tech, Colette also worked on issues relating to agriculture and natural resource management, especially developing innovative approaches to extension and transformatory learning methods for community development, and on issues around aflatoxins in East Africa. Her main geographical region of expertise is Central Asia, particularly Tajikistan. She has also worked in other parts of Asia, Latin America and East and West Africa. Among her publications are two monographs – *Control and subversion: Gender relations in Tajikistan*; and

Muslim youth: tensions and transitions in Tajikistan. She has also written on extension methodologies, aflatoxins, and ways in which gender power relations contribute to poverty. Colette has a PhD in social sciences from the Universiteit van Amsterdam in the Netherlands.

Carmen Suarez-Capello (PhD) is an Equadorian Agricultural Engineer specializing in crop protection. She has more than 30 years agricultural research experience, mainly in the area of developing integrated pest and disease management technologies for plantain and cocoa. During the last 10 years, she has been involved in a participatory research and technology transfer initiative which has generated innovative methodologies for teaching farmers complex biological integrated pest management concepts. She has authored many scientific articles, book chapters and other publications for farmers and extension agents in journals and proceedings of international meetings. Among her publications are:

"The pathosystem". *In Management of witches broom in cacao; Monograph of the International Witches broom project.* Longman, U.K.

Reproducing witches broom outside cocoa producing countries. PANS, England. 26(4): p.p. 435-436.

Narisi Mubangizi holds a Masters degree in Agricultural Extension Education from Makerere University. For the past four years he has been engaged in community-level agricultural extension service delivery with various non-governmental organizations. He is currently employed by Kabale Catholic Diocese as the coordinator of a sustainable agriculture program. Among his publications are:

"Information sources and constraints of private service providers under the national agricultural advisory services in Uganda: A case study of Arua and Tororo districts in Uganda" *Uganda Journal of Agricultural Sciences, Vol. 9 (3);*

Challenges and opportunities of private agricultural extension service providers in accessing and utilizing agricultural information under the National Agricultural Advisory Services in Uganda: A case study of Arua and Tororo districts in Uganda, The XIth Biennial IAALD/USAIN

(International Association of Agricultural Information Specialists/United States Agricultural Information Network) conference held in Lexington Kentucky, USA 15-19 May 2005.

Richard Miiro has a Masters degree in Agricultural Extension Education from Makerere University. He has over 10 years experience as a lecturer in the Department of Agricultural Extension/Education, Makerere University. He has been involved in urban and rural agricultural research and development projects in the area of extension systems, community-based agricultural technology transfer and farmer learning, action research, and organizational and institutional systems.

Orum Emuria Boniface has a masters degree in Agricultural Extension Education from Makerere University. He has over five years of teaching experience as an assistant lecturer in the Department of Agricultural Extension/Education, Makerere University and over six year's field experience in rural extension, agricultural development, community development and food security, working with international and local non-governmental organizations.

Francis Byekwaso (PhD) is the manager, Planning Monitoring and Evaluation, of Uganda's National Agricultural Advisory Services.

Vincent Kayanja has a Masters degree from Wageningen University in the Netherlands. He is the National Agricultural Advisory Services coordinator, Kasawo Sub-country, Mukono District, Uganda.

Frank Matsiko Biryabaho (PhD) has over 15 years experience as a lecturer in the Department of Agricultural Extension/Education, Makerere University. He has also worked as a consultant to various international and national organizations.

Foreword

This book makes a vital contribution to the documentation and analysis of Uganda's experiences with agricultural extension systems. Introduced way back in the early 1900s by the colonial administration, agricultural extension has since been part and parcel of Uganda's agricultural development strategy, albeit with various changes and innovations in approaches, institutional arrangements, methods, and strategies. Notable innovations included liberalization in the early1990s, which broadened the provision of extension services beyond the public domain to include a range of non-governmental organizations (NGOs); decentralization of extension services to the districts in 1997 (Local Government Act, 1997); and the shift from public to private farmer-owned contract extension approach through the National Agricultural Advisory Services (NAADS) Act, 2001. While all these provided a wide range of valuable experiences, systematic documentation and analysis with a view to drawing lessons to inform theory and practice, has been largely insufficient. Unfortunate consequences of this dearth of information include, among others, repetition of previous mistakes, policy reviews not informed by previous lessons, localized success stories not widely disseminated and up-scaled, and the absence of credible locally-produced reference text books for use during training of agricultural extension professionals at university and other levels.

This book should serve as an important resource for educators and students in agricultural higher education, policy makers and practitioners in Uganda and other developing countries. The authors of the chapters in this book have written from first-hand knowledge and experience gained from serving in the agricultural extension education system as educators, managers and field practitioners in Uganda and elsewhere. I appreciate the contribution of Colette Harris and Carmen Suarez-Capello whose experiences in Ecuador reflects new thinking that has the potential to add value to extension methodological innovations in Uganda and elsewhere. It is my belief that this volume is a valuable addition to agricultural extension literature.

Hon. Eng. Hilary Onek

Minister of Agriculture, Animal Industry and Fisheries, Uganda

Acknowledgement

This volume is published as part of a Rockefeller Foundation funded project titled *Building capacity for training and research of the Department of Agricultural Extension/Education, Makerere University*. Special appreciation is therefore extended to the Rockefeller Foundation for the financial support, and in particular, Dr Ruben Puentes, for encouragement and support in securing the funding.

The volume represents a collective effort by the entire Department of Agricultural Extension/Education, Makerere University. I therefore appreciate the institutional support of the Department as well as the individual staff for their enthusiastic participation in writing the proposal to raise the funds, and constructive feedback on the manuscript.

My gratitude is also extended to the contributing authors. Their appreciation of the volume's subject and tailoring their chapters to the overall theme of the volume makes for greater in-depth appreciation of the issues and challenges confronting agricultural extension in Uganda as well as potential and tried innovations and lessons learnt.

Appreciation is also extended to Dr Frank Matsiko Biryabaho, acting head, Department of Agricultural Extension/Education, Makerere University; Dr. Jeff Mutimba, regional coordinator for the Sasakawa Africa Fund for Extension Education; Dr William Epeju, lecturer, Kyambogo University; Mr Vincent Kayanja, Sub-county NAADS coordinator, Mukono District and Dr Joseph Oryokot, technical services manager of the National Agricultural Advisory Services, who served as critical reviewers for selected chapters.

Dr Margaret Najjingo Mangheni

Editor

CHAPTER 1

Introduction

Margaret Najjingo Mangheni

Uganda is a developing country whose economy is predominantly agrarian, with 36% of the gross domestic product, 81% of the employed labour force, and 31% of export earnings derived from the agricultural sector (http://w ww.nationsencyclopedia.com/africa/Uganda, 3/28/2007). The agricultural sector, consisting of smallholder farmers cultivating an average of about 2.5 acres each, largely involves subsistence farming with about 70% of the area under cultivation being used to produce locally consumed food crops. Despite the sizeable investment of resources into the sector, the majority of farmers produce with little or no utilization of scientific technological advances. It is undisputable that agricultural extension in Uganda, which is supposed to perform the role of catalyzing farmer learning, access and utilisation of information and technologies, as well as coordinate the multi-actor innovation system, is yet to have the anticipated impact. Looking back over its history, one sees a catalogue of experiments with approaches and methodologies. What is glaringly lacking, though, is a systematic analysis of the processes, and an attempt to clearly link new approaches to previous ones, drawing on lessons learnt. Instead, the changes seem to reflect donor interests (such as the introduction of the Train and Visit approach in 1993), and/or global trends (rather than local needs or reality), as was the case with the privatized contract extension approach introduced in 2001. While it is important to keep in step with global trends, responses must reflect local reality, which in this case includes insufficient relevant capacities amongst the private sector, farmers and other players; diversified farming systems; weak farmer institutions; political dynamics and limited resources, among others, so that necessary adaptations can be made.

Agricultural extension has consistently been one of the key policy strategies used to bring about agricultural transformation in Uganda since its introduction by the colonial administration in the late 1800s. Semana (1999) clustered the history of Uganda's extension service into five eras, namely, 1898-1956, 1956-1963, 1964-1971, 1972-1980, and 1981-1991. The early period (1898-1956) was characterized by emphasis on distribution of planting materials for major cash crops for export and simple messages on how to grow those crops; implementation by chiefs, a few expatriate field officers and African instructors; coupled with enforcement of bye-laws requiring households to follow specified agricultural practices such as soil conservation, and having adequate food stocks. During the period 1956-1963, the extension strategy concentrated on progressive farmers who were provided with technical advice, inputs and credit, with the expectation that their improved performance would have a demonstration effect on the rest of the community, attracting others to learn from them. In the period 1964-1971, the extension service adopted a more professional approach, with the role of extension workers becoming more educational. This however was reversed during the period 1971-1980, which was characterized by disruption of the economy and service delivery, greatly reducing the effectiveness of extension. Staff concentrated on selling inputs to the detriment of their educational role. Semana (1999) refers to the subsequent period from 1981 to 1991 as the recovery phase, with a strong focus on staff training, better linkages with research, farmers, NGOs and other institutions. The period from 1992 to 2001 saw several reforms with direct implications for relevance and effectiveness of agricultural extension services in the country, major ones being liberalization and privatization, which attracted a multiplicity of extension service providers to supplement government extension; decentralization, civil service reform accompanied by retrenchment of staff; and restructuring/unification of agricultural sector ministries. However, by and large, all extension activities have tended to be organized around short-term, localized projects targeting specific needs and client groups in selected geographical areas such as Extension Saturation project; Young Farmers of Uganda; mechanization; credit; and cotton projects; the Agricultural Development Project in Tororo, Kumi, Lira, Soroti, Kitgum and Gulu; the Agricultural Rehabilitation project in Bushenyi, Mbarara, Rukungiri, Kabale and Kisoro; and Agricultural Extension Project. These however have not been guided by a coherent agricultural extension policy, which probably explains the haphazard manner that has characterised the bulk of the interventions.

A critical analysis of Uganda's agricultural extension history reveals a number of issues and lessons. A common thread running throughout the entire period from its inception to the 1900s is a tendency to serve the interests of governments rather than farmers. For instance, while the approach adopted by the colonial government was suitable for introduction of new crops because farmers were assured of planting materials, training on basic agronomic practices and markets, an inherent weakness was that the approach was primarily meant to serve the interests of the colonial government, not the farmers, by focusing on selected crops required by the former; and using force to ensure adoption rather than education, which undermined the sustainability of the practices. On the other hand, while the strategy of targeting contact/progressive farmers was suited to the limited funds and staff numbers at the time, it failed to achieve the desired coverage because, in some instances, the selected progressive farmers were not willing to share information with others, while in other cases the ordinary farmers perceived the progressive ones as a privileged group whom they could not emulate. By and large, effectiveness and sustainability were, among other things, undermined by a lack of resources particularly at the lower levels, poor supervision and deterioration of staff morale. Organizing services around commodities and projects meant that services were localized, short term, and focused on a limited range of farmer needs. Furthermore, agricultural extension in Uganda has largely depended on external funding from donors. Although the government consistently pays staff salaries, funds to facilitate their operations in the field are often lacking, such that any meaningful extension activity only happened under donor-funded projects. Insufficient government commitment to providing resources on a consistent basis undermined continuity. In addition, external forces, particularly from donors, tended to drive the changes in approach, hence limited local ownership.

The reforms introduced in 2001, involved privatization of extension services. They were an attempt to respond to prevailing global trends and national challenges facing agricultural extension. Rivera and Zijp (2002), attributes this to a change in the world's ideology characterized by a power shift from public sector services to private sector hegemony, a transition towards global capitalism and free market principles. Accordingly, the new ideology, has led to developing countries being pressured to reform their public sector systems. Public sector agricultural extension is increasingly pressed to adopt various decentralization, cost-recovery and privatization strategies (Rivera and Gustafson 1991; Smith 1997). One notable development in this era of change has been that of the public sector contracting out the delivery of extension services (Rivera and Zijp, 2002).

In Uganda, these trends were in response to the various challenges confronting agricultural extension, namely, inefficiency and ineffectiveness of bureaucratic government ministries; supply-driven top-down approaches that serve the interests of external constituencies other than that of farmers; dwindling public resource allocation to agriculture in general and agricultural extension in particular; the need to move from an emphasis on technology transfer per se to poverty alleviation and enhanced rural livelihoods as a goal for agricultural extension interventions; as well as emphasis on crosscutting issues such as environmental sustainability and equitable access to services by various vulnerable groups, such as those affected by HIV/AIDS and women.

Organization of the Book

The chapters in this volume contain a wealth of experiences from selected agricultural extension initiatives in Uganda and elsewhere, with valuable lessons and insights for addressing challenges facing agricultural extension in Uganda and other developing countries. The book is organized into four sections. Section I contains two chapters focusing on Uganda's experiences with a privatised contract farmer-owned extension approach introduced in 2001. The two chapters in Section II focus on experiences of two innovative extension methodologies used in Uganda and Ecuador. Section III has two chapters on training and capacity development for extension innovations and reforms. Section IV contains chapters on emerging issues that are assuming importance in agricultural extension service delivery, namely, the implications of HIV/AIDS and the environment on agricultural extension.

Section I: Uganda's Experiences with a Privatised Contract Farmer-owned Extension Approach

In 2001, Uganda embarked on a process of transforming its public extension system in conformity with the rest of its economic transformations. The public extension system was gradually phased out and replaced by a contract privatized system. It was implemented by the National Agricultural Advisory Services (NAADS) within a broader policy framework of the multi-sectoral Plan for Modernization of Agriculture (PMA) involving decentralization, liberalization and privatization. In operational terms, this means that the private advisory service providers operate on a contract basis with farmer organizations. The major features of the NAADS program include private delivery by means of public funding; demand-driven and farmer-owned; decentralized service delivery and poverty and gender targeting. Chapter Two identifies key requirements for success of the privatised contract farmer-owned agricultural extension approach, based on an analysis of Uganda's experiences. The chapter adopts a global perspective, blending the NAADS experience with others elsewhere in the world so as to generate general recommendations for policy-makers and practitioners. The requirements for success include a conducive policy environment, sufficient farmer capacity to articulate their demands, sufficient private service-provision capacity, efficient and effective service quality assurance mechanisms, adequate and sustainable funding, conditions conducive to profitable private service provision, and effective coordination of the multi-actor processes that are part and parcel of this complex approach. The chapter makes recommendations for policy-makers and practitioners in order to contribute to further thinking and refining of the evolving practice.

Chapter Three analyses the NAADS program's experiences regarding implementation of participatory monitoring and evaluation (PME) in selected districts. PME provides a systematic process for self-reflection and learning, documenting experiences and lessons. Such a system has the potential to empower local people to demand greater social responsiveness and ethical responsibility, within communities or groups.

Benefits realized from the practical application of the PME approach are described namely, improved communication among stakeholders, improved planning, strengthened quality assurance of services, enhanced farmer empowerment and increased reliability of monitoring and evaluation data. The chapter also shares lessons learnt by implementers during the process. PME development is slow and evolutionary and requires continuous adaptation; it requires trained, committed and stable staff as well as supportive structures and institutions; it functions more effectively with farmers who are already empowered; and it needs to be part and parcel of an overall program monitoring and evaluation framework. For PME to be successful it requires a competent technical team at district and sub-county levels charged with overseeing the process. The chapter finally discusses a range of challenges that were experienced, among which are poor quality information generated by farmers, delayed reporting, weak feedback mechanisms, poor attendance by farmers of group meetings and the multiplicity of local languages.

Section II : Innovative Extension Methodologies

The cases presented in this section depict practical experiences with innovative ways of transferring technology and facilitating farmer learning. In Chapter Four, a methodology for reaching urban farmers is presented. Despite its significant contribution to increasing household livelihoods, particularly among the urban poor, urban agriculture has received limited support in many countries, including Uganda. In most cities, both crop and livestock farming are not considered a legal activity. In Kampala, providing extension and advisory services to urban farmers hardly covers the full range of needs except for livestock and poultry for which farmers seek support from Veterinary personnel. One of the factors contributing to inadequate urban agricultural services is the dearth of methodologies suited to this unique context. Providing extension services in cities demands innovative methodologies to enhance farming activities in line with city regulations. Chapter Four presents the case of an innovative technology dissemination approach used to address a food and nutritional security need in Kampala city, Uganda, through the Orange-fleshed Sweet Potato project. The approach used schools as technology learning and dissemination centers, farmer to farmer extension approaches and an array of other extension methods to reach farmers. It was based on the extension principle that the use of multiple methods to reinforce each other results in greater levels of learning and technology adoption. Using school gardens to transfer practical agriculture knowledge to pupils and students took advantage of government of Uganda policy that requires all pupils in primary schools to take the subject of agriculture because of its critical role in the country's economy. The chapter describes the process, challenges experienced and lessons learnt.

Chapter Five focuses on experienced transformative approach to extension for plantains on the Ecuadorian coast. It represents a departure from the general tone of the other chapters in that it is based on an experience outside Uganda. The chapter questions the often-used top-down, simplistic approaches, such as lectures and occasional field days during which extension agents demonstrate certain techniques, show farmers how to apply them, but fail to take farmers' responses or ideas into account. Moreover, even when participatory methodologies are used, extension agents are seldom trained to facilitate in such a way that farmers develop their own learning skills. Instead, agents tend to explain what to do and simply repeat it until the farmers understand it well enough to use it for themselves, rather than helping farmers grasp the principles behind the farming techniques. The problem with this approach is that, if the agro-ecosystem environment changes even a little, the technologies may fail, leaving farmers without ways to cope with the change. In other words, the use of participatory methodologies has failed to make significant improvements in adoption of new technologies.

Chapter Five defines and analyses the strengths and limitations of the learner-centred Farmer Field School (FFS) methodology, drawing on experiences in various countries. A novel transformative extension approach developed for and tested among plantain farmers in Ecuador is presented as an improvement on the FFS. It uses photographs creatively in conjunction with farmers and other local materials to provide a similar level of exploration of the agro-ecosystem as in FFS weekly group observations, thereby encouraging farmers to work through the causes and effects of processes by themselves, to perceive relationships in situations where these are not apparent, rather than *telling* farmers what they should know. Its basic tenets include experiential learning, a discovery-based approach, and incorporating farmers' ideas in developing extension materials. The authors adopt a practical step-by-step approach in presenting the actual modules to facilitate farmer experimentation, exploration and discovery of various issues related to plantain management in a way that is appealing to practitioners who may wish to adapt the methodology in other settings.

Section III: Training and Capacity Development for Innovation and Reform in Agricultural Extension

In Chapter Six, the educational implications of the policy shift from a public to a private contract agricultural extension system are analyzed. The change calls for new relationships, knowledge and skills among the key stakeholders from the private sector, farmers, farmers' organizations and the government. These new needs create new demands on the agricultural education system, not only in terms of appropriate curricula but also regarding instruction and training delivery approaches. The chapter presents results of a series of studies conducted over a period of fours years between 2001 and 2004, on perceptions among stakeholders, including private contracted agricultural advisory service providers, on the new functions, challenges and agricultural education needs brought about by the privatized contract agricultural advisory system. On the basis of the analysis, the chapter presents implications for design and implementation of educational programs to meet the diverse needs arising from this system adequately.

Chapter Seven addresses the challenge of responding to demand for agricultural training by public universities which often have rather rigid bureaucratic structures unsuited to dynamic, changing environments that call for timely interventions. With transformations in agricultural policy characterized by market-oriented private-sector driven approaches, the traditional university culture is confronted by a challenge if it is to service the new privatized extension system. A critical pre-condition for the success of the transformation process is the existence of a pool of well-trained advisory services providers to drive the agricultural modernization process, and universities like Makerere are better positioned to contribute towards this pool. The chapter analyzes Makerere University's experiences with a demand-led mid-career Bachelor of Agricultural Extension Education degree program and recommends changes that universities should make if they are to offer demand-led need-based innovative programs.

Section IV: Selected Emerging Issues in Agricultural Extension Service Delivery

Literature is rife with challenges and issues confronting agricultural extension today. This volume addresses two of these issues.

Chapter Eight tackles the issue of the relationship between agricultural development and the environment. While agriculture continues to be a major source of livelihood for the majority of people in Uganda, the sector remains one of the major cause of environmental degradation. Environmental degradation arising from the inappropriate use of agricultural practices in turn limits agricultural production and productivity. Consequently, agriculture becomes both a cause and a victim of environmental degradation. It is therefore impossible to talk about modernization of agriculture and eradication of poverty without talking about sustainable use of natural assets and ecosystems that underpin agriculture. The chapter highlights the effects of agriculture on various components of the environment, namely, soils, biomass, wetlands, rangelands, water resources, climate, social harmony and the general atmosphere, drawing implications for agricultural extension. It illustrates existing opportunities and challenges for integrating environmental messages in extension; and the role of extension in promoting sustainable use of natural resources for agricultural development.

Another issue confronting agricultural development is the HIV/AIDS pandemic. Uganda has been affected by the pandemic for almost a quarter of a century, with far-reaching social and economic consequences that have affected individuals of all walks of life and communities nationwide. The impact of the disease has been mainly in increasing levels of morbidity and mortality that disproportionately affect women and men. For Uganda, food insecurity, degraded livelihoods, increased vulnerability and adverse socio-economic impacts have been identified as main causes and consequences of HIV and AIDS. At community level, declining productivity in agriculture is eminent and the death of prime age adults has imposed unsustainable strains on the extended family structure due the massive burden of orphans. Nowadays an increasing number of households are headed by children. It is becoming increasingly evident, therefore, that the impacts of HIV and AIDS are undermining development initiatives.

Given the critical role of the agricultural sector in national development and the livelihoods of the majority of Ugandans, impacts of HIV/AIDS on the agricultural sector and agriculture based livelihoods have far-reaching implications. Chapter Nine discusses the impacts of the pandemic on farming systems, livelihoods of rural households, social institutions, and the agricultural estates sector. It examines the role and challenges of agricultural extension in responding to the epidemic.

References

Rivera, W.M. and Zijp, W. (eds.) (2002): *Contracting for Agricultural Extension: International Case Studies and Emerging Practices.* New York: CABI Publishing.

Rivera, W.M. and Gustafson, D.J. (eds.) (1991): *Agricultural Extension: Worldwide Institutional Evolution and Forces for Change.* Amsterdam:Elsevier.

Semana, A.R. (1999): "Experiences of Agricultural Extension in the 100 Years of Agricultural Research in Uganda". Unpublished paper.

Smith, (1997): "Decentralization and Rural Development. Rome: FAO/SARD. Local Government Act (1997).

http://www.nationsencyclopedia.com/africa/uganda, 3/28/2007

SECTION I:
UGANDA'S EXPERIENCES WITH A PRIVATISED CONTRACT FARMER-OWNED EXTENSION APPROACH

.

CHAPTER 2

Requirements for Successful Privatised Contract Farmer-led Agricultural Extension:
Lessons from Uganda's National Agricultural Advisory Services

Margaret Najjingo Mangheni and Narisi Mubangizi

Introduction

Agricultural extension is organized in different ways to address specific problems and accomplish a variety of objectives. Therefore, various agricultural extension systems and approaches have been used in Uganda and other countries. An agricultural extension approach refers to how extension services are organized and delivered; and therefore consists of the goals/objectives, target clientele, organizational features, delivery methods and techniques, as well as the philosophy that underpins all actions geared towards discharging its functions.

In the past, in many countries, including Uganda, the two basic components of agricultural extension systems, namely funding and delivery, were handled by the public sector. However, a shift towards greater private sector participation in the provision of extension services is being experienced globally (Rivera, 1991; Rivera and Zijp, 2002). This shift is attributed to the perceived ineffectiveness, irrelevancy and irresponsiveness of public extension services, and budgetary constraints, especially in developing countries (Rivera, 1991; Rivera, Zijp and Alex 2000; Rivera and Zijp, 2002) which has led to the emergence of various arrangements. Countries, especially in the developing world are at different levels of modifying their extension systems, moving away from public-sector dominated systems. Rivera and Cary (1997) list four possible arrangements for the provision of extension services, some of which are already being applied in some countries.

- Both funding and delivery are by the public sector. This has been the commonest mode of providing extension services in developing countries. It is seen to be a suitable arrangement for poor categories of farmers and for those enterprises that might not be very profitable but important for achieving goals of public good, such as social equity and environmental conservation.

- Funding and delivery are both by the private sector, as exemplified by extension services by farmers' associations, agro-inputs and commodity-based and other commercial companies.

- Funding is by the private sector and delivery is by the public sector, examples being fee-based services in Mexico and New Zealand (Rivera and Cary, 1995) and facilitation of public extension staff by NGOs, as observed in Uganda (Rivera et al., 2000).

- Funding is by the public sector and delivery is by the private sector, for example the privatized contract farmer-led extension by Uganda's National Agricultural Advisory Services (NAADS).

This chapter analyses and draws lessons from the utilization of the privatized contract farmer-led agricultural extension approach of NAADS, with a view to identifying key requirements for successful application of the approach. The chapter adopts a broad and, to some extent, global perspective, blending the NAADS experience with others elsewhere in the world so as to generate general recommendations for policy-makers and practitioners.

An Overview of the Privatized Contract Farmer-led Agricultural Extension Approach in Uganda

Uganda's privatized contract farmer-led agricultural extension approach started in 2001 after an act of parliament, the NAADS Act, was promulgated. The premise was that the previous agricultural extension system had failed to bring about greater productivity and expansion of agriculture despite costly government interventions. The program began with a trailblazing phase in six districts and spread to other districts throughout the country.

Goals of the NAADS Program

The vision of the NAADS program is a decentralized farmer-owned and private-sector serviced extension system that contributes to the commercialization of agriculture and consequent poverty eradication. The program therefore aims principally to develop a demand-driven, client-oriented and farmer-led agricultural service delivery system. In so doing, NAADS emphasises farmer empowerment through farmer institutional development, capacity building and use of participatory approaches.

Target Farmers

NAADS targets different categories of farmers, the majority of whom are due to the nature of Uganda's agriculture, small-scale subsistence farmers. It targets principally economically active poor farmers with limited capacity (in terms of land and other assets) Within this group, emphasis is placed on women and other marginalized categories of people through gender mainstreaming of all its operations. However, operationally, effective targeting and involvement of economically destitute people (those without any assets) has been rather challenging and is yet to yield the desired results.

Institutional Arrangement

The NAADS is an autonomous program within the Ministry of Agriculture Animal Industry and Fisheries. It is coordinated by a small national secretariat located in the capital city of Kampala, which is mainly responsible for overall planning, coordination and linkage with the various stakeholders. At the secretariat, overall program oversight is the responsibility of the Board, while day-to-day implementation is by the technical staff, consisting of an executive director and staff in the departments of technical services, planning, monitoring and evaluation and finance and administration. Below the national secretariat, NAADS uses the existing decentralized local government structures; with district and sub-county coordinators in each of the districts and the sub-counties of operation respectively.

In order to enhance farmer ownership and control over the extension system, NAADS, through the farmer institution development process, facilitates establishment of district-level farmers' fora, sub-county farmers' fora, enterprise-based farmers' groups and, more recently, parish coordination committees[1]. All these farmers' institutions are elected by farmers themselves and have defined terms of reference. However, the most critical of these is the sub-county level farmers' forum which aggregates the service demands of all the groups in the sub-county, to formulate the demands that form the basis for contracting private service providers. In addition, the sub-county level farmers' forum contracts service providers and monitors their performance to ensure that they deliver the agreed services. In executing their duties, the sub-county farmers' forum is technically supported by the sub-county technical committee, composed of the sub-county NAADS coordinator, the sub-county chief and the technical audit team at district level, which periodically visits sub-counties to audit the quality of the services delivered by the contracted service providers. The role of individual farmer-group members and the parish coordination committees is to observe what happens in their areas of jurisdiction and to inform the sub-county level farmers' forum to take action as necessary. Each of these levels of farmers' institutions is supported by NAADS in terms of capacity building through farmer institutional development, to enable them perform their assigned tasks.

Role of Extension Personnel (Service Providers)

Private extension service providers (who operate either as individuals or firms), are contracted by sub-county farmers' fora to deliver specific services, which normally constitute enterprise-specific advice to specific farmers' groups over a period of three to six months. The service providers are normally given terms of reference that spell out what specific practices they are supposed to train and advise farmers on, and what type and number of demonstrations they are expected to establish. The terms of reference therefore form the basis for service providers' work. The role of service providers is largely limited to training and providing advice to individual farmers on topics ranging from agronomic practices through processing to marketing. There are, however, service providers (mainly NGOs government organizations and community-based organizations CBOs) who, through memoranda of understanding, are engaged in training and supporting farmers' groups to develop into strong and sustainable institutions to enable the smooth implementation of the NAADS program.

---

[1]   In Uganda, the country is divided into districts, sub-counties, parishes and villages, according to the Local Government Act, 1997.

Extension Needs Assessment

Farmers' needs regarding advisory services form the basis of service providers' work. Farmers organized into groups at village level are guided by NGOs or CBOs during farmer institutional development to select priority enterprises (normally a maximum of three enterprises per group). The process involves the use of criteria generated by the NAADS secretariat (emphasizing marketability and profitability) to score and rank the enterprises.[2] The selected priority enterprises are analyzed further by the groups in terms of their constraints and opportunities. The list of enterprises and the associated constraints and opportunities are aggregated at parish and sub-county level to form the sub-county priority list, which then forms the basis for private service-provider recruitment.

Requirements for Successful Privatized Contract Farmer-led Extension

This section analyzes the design and operationalization of the privatized contract extension approach in Uganda with a view to identifying gaps and lessons. The NAADS experience and lessons from elsewhere in the world reveal the following essential requirements for successful privatized contract farmer-led extension: (a) a sound policy environment for agricultural development (Rivera and Zijp, 2002); (b) sufficient farmer capacity to articulate their demands and service-quality requirements (Garforth, *et al*, 2003), as well as empowerment to supervise service providers (Rivera *et al*, 2000; Garforth, Angell, Archer and Green, 2003; Chapman and Trip, 2004); (c) sufficient private service-provision capacity to satisfy the demand in terms of quality and quantity; (d) efficient and effective mechanisms for ensuring adherence to minimum service quality standards (Rivera *et al*, 2000; Rivera and Zijp, 2002); (e) adequate and sustainable funding; (f) conditions conducive to profitable private-service provision. Lastly, given the multi-actor nature of this approach, the need for clarity regarding institutional roles, participation by all and effective coordination cannot be over-emphasized. Each of these factors is discussed in depth below.

---

2   The enterprise selection criteria with their weighting in brackets are as follows: profitability of the enterprise (4), availability of market (3), low financial outlay (2), low risk (2), and farmers' prior knowledge about the enterprise (1).

Sound policy environment for agricultural development

A sound policy environment for agricultural development, with ample provisions for all agricultural support services, including farmers' access to inputs, markets and suitable microfinance, is basic to any meaningful extension service (Rivera and Zijp, 2002). Uganda's privatized contract farmer-led agricultural extension approach has been implemented within a policy framework of liberalization, privatization and decentralization, with the overall goal of improving the livelihoods of the population. In order to eradicate high levels of poverty, the government designed a Poverty Eradication Action Plan (PEAP) in 1995. It was launched in 1997 and revised in 2000 and 2004. The PEAP advocates poverty eradication through agricultural modernization, employment creation and industrialization. Under this umbrella falls the Plan for Modernization of Agriculture (PMA), which represents the country's comprehensive multi-sectoral framework for eradicating poverty by transforming the livelihoods of farmers from subsistence to commercial (MAAIF, 2000). Agricultural advisory services are one of the seven principal pillars of the PMA, others being research and technology development, agricultural education, rural financial services, agro-processing and marketing, sustainable natural resource utilization and management, and supportive physical infrastructure. However, it is worth noting that implementation of the other pillars has lagged behind, with the resulting challenge of advisory services being offered without the other, necessary agricultural development services being available. Farmers therefore fail to make optimum use of technologies and information to better their livelihoods. The potential impact of the program has most likely been affected significantly by this vacuum.

Farmers' capacity to articulate their demands

For a demand-driven, farmer-controlled extension system to work effectively, farmers should be able to articulate their needs clearly (Garforth *et al.*, 2003). The clarity and specificity of farmers' demands is especially critical given that the demands form the basis for the selection, contracting and evaluation of private service providers. In situations such as those prevailing in Uganda, where it would not be cost effective or efficient to target individuals, farmers need to be organized in a way that fosters collective articulation of demands in an all-inclusive manner. Success would therefore hinge on the effectiveness of the farmer institution development process in developing cohesive, sustainable groups, coupled with a demand articulation process that identifies farmers' true or real needs.

In this regard, the NAADS has, since its inception, placed emphasis on farmer institutional development (FID) (NAADS, 2005). FID activities include community-level mobilization, farmer-group formation and registration (at village level and beyond), training of groups, election and training of farmer-forum members, participatory selection of agro-enterprises, farm technology development and promoting group and individual savings. As a result of these FID activities, 20, 000 farmers' groups with nearly 400, 000 members had participated in NAADS activities in the first four years of implementation (NAADS, 2005). However, many of these groups were externally motivated and lacked the basic ingredients for sustainability. The problem is more pronounced in certain areas, such as central Uganda, that do not have a strong culture and/or history of collective action. The demand articulation process also has inherent limitations.

Though the process of setting the advisory service agenda at sub-county level involves a high level of farmer participation, it tends to be prescriptive. Obaa, Mutimba and Semana (2005) based on an in-depth study of the NAADS enterprise selection process in Mukono district, noted that the establishment of conditions such as the number of enterprises to be selected, selection criteria, the administrative unit to be covered (sub-county), and the farmers who could participate in the process by the NAADS secretariat were all indicators of the rigidity of the process. Obaa *et al.* further note that some of the criteria established for enterprise selection and advisory-service needs assessment were too mathematical and complicated for the farmers' competence levels, and as a result excluded women and the poor, whose education levels are often low. Similar conclusions were reached by Draa (2005) in a case study of Tororo and Arua districts. Draa noted that most farmers failed to understand the NAADS demand assessment process. Consequently some, whose priority enterprises did not feature in the final list, felt that the program was imposing priorities on them.

Farmer's Capacity to Articulate Service-Quality Requirements and their Empowerment to Supervise Service Providers

Boettcher, Collenberg, Huppert, Keller, Mack, Pilgram, Schutz, Springer-Heinze and Weiskopf (1999) note that Germany's contract extension system's success depended mainly on the quality-control capacity of clients and a strict service-provider qualification certification mechanism. Rivera *et al.* (2000) and Chapman and Trip (2004) also note that farmers have to be able to monitor and evaluate the services received if a contract farmer led extension system is to succeed. These experiences concur with lessons from the NAADS Program.

Mubangizi (2006), in a case study on needs and client responsiveness of private service providers under NAADS, conducted between 2003 to 2004 in Arua and Tororo districts, noted that farmers were generally able to specify their advisory-service quality requirements. They could articulate the desired advisory-service quality attributes related to training, demonstrations, and personal behaviour and competence of service providers. Interestingly, the advisory-service quality attributes generated by the farmers were closely related to those generated by the NAADS quality assurance team through consultations with various stakeholders in Uganda in 2003. Mubangizi further found that, not only were farmers able to articulate their advisory-service quality requirements, they were also, to some extent able to assess objectively service providers' performance on such requirements. However, the existing monitoring and evaluation procedures, which relied on the technical team and the farmers' fora at sub-county and district level, failed to tap farmers' capacities in this area. Since the process was often conducted outside the training environment, ordinary farmers' input was seldom sought. Accordingly, the farmers' forum became more aware and confident about their role in supervising and controlling service providers, while the majority of farmers at group level (including group leaders) remained unaware of the expected service standards. Farmers' lack of awareness about their roles in service-quality assurance seems to be compromising quality and their satisfaction with the services. However, in communities where participatory monitoring and evaluation has taken root, such tendencies are on the decline (see Chapter Three of this volume).

Private Service Provision Capacity

An adequate and well-functioning private agricultural extension service-provision sector is an essential prerequisite for an effective response to farmers' advisory-service demands (Rivera and Zijp, 2002). The NAADS experience revealed that private service providers require capacity in the area of financial resources, transport facilities, information-processing equipment and technical capacity in the requisite competence areas. In this case, evidence of weak private service-provision capacity included the inability of most private service providers, especially individuals, to afford any reasonable investment in basic equipment and infrastructure, such as transport, information-processing equipment and office space (Mubangizi, 2006). Those private service providers operating as individuals were found to be much more constrained than firms. There were also instances (particularly in the early stages of the program) of districts failing to attract sufficient applicants willing to offer their services to certain enterprises. The required competencies were often lacking, since many underwent training to prepare them for the public-sector work environment.

A number of barriers to private service-provision capacity development have been identified by Rivera *et al.* (2000), and they include bureaucracy in government, limited career opportunities and prospects for professional growth, lack of purchasing power for most small-scale farmers and inadequate rural infrastructure, that results in high costs of operation and reduced geographical coverage, among others. Most of these barriers exist in Uganda and have been aggravated by the absence of a clear and comprehensive strategy for private service-provider capacity development under the program, coupled with absence of institutionalized linkages with research and education institutions that would otherwise provide essential technical support to service providers. The existing guidelines are silent about the role(s) of these institutions, let alone how these roles could be executed effectively. Proper definition of the procedures and roles of each of these stakeholders is critical if the role(s) of each are to be realized with minimum conflict of interests and duplication of efforts. As Rivera and Zijp (2002) correctly point out, one of the key roles of government in privatized extension systems is to provide training and technical information to the agencies contracted.

Service-quality Assurance Mechanisms

According to Rivera *et al.* (2000), in addition to the existence of an adequate private service-provision capacity, mechanisms for ensuring adherence to minimum advisory-service quality standards are crucial for the success of the contract extension system. Emphasis should be on the quality of technology passed on to farmers and the way farmers are guided to use it to obtain desirable, sustainable increments in productivity and income. This calls for a quality-assurance mechanism that evaluates the quality of the providers of the technology, the technology itself, and the processes of providing and utilizing the technology to transform the agricultural system. However, Mubangizi (2006) found that the existing quality-assurance mechanisms under NAADS evaluated the quality of the service providers' qualifications by requiring service providers to register at district level, and assessing them through a competitive process that considers formal education qualifications and work experience, among other attributes, before awarding contracts to service providers. In addition, quality assurance of the process of advisory-service provision was done through checks by the technical audit team, normally at the end of the contract period, and by monitoring visits by the sub-county farmers' forum and local leaders. There was a limited attempt to ensure the quality of the advisory services (especially information) before delivery to farmers.

The effectiveness of the above-mentioned practices have not been determined, but existing reports present a mixed picture. Technical audits have apparently been found to be ineffective in ensuring service quality (NAADS, 2005; Mubangizi, 2006). Mubangizi attributed the limited effectiveness of the technical audits to their inappropriate timing and mode of conduct. Technical auditing exercises were normally done in the middle or towards the end of the contract period, when the information (whether accurate or not) had already been passed on to the farmers. In addition, some of the service providers being audited were never informed of when and how the technical audit was to be carried out and consequently never participated. Both of these scenarios compromise the ability of the technical audits to foster collective learning and corrective action, which is essential for effective service delivery.

A number of other factors also cast doubt on the effectiveness of the service-provider quality-assurance mechanisms. Firstly is the issue of limited capacity among existing technical local government staff to undertake quality assurance effectively given their increased workload and limited staff numbers. Secondly, farmers had little information about service providers' personal profiles such as qualifications and experience, among others, (Mubangizi, 2006), which could open the door to unregistered persons. Indeed, there were reported cases of firms using trainers other than those whose curricula vitae had been presented to win the contract. Thirdly, corruption at the sub-county level during the process of contract award compromised the quality of the service providers, since best service providers were not necessarily awarded the contracts. The need among service-providers to compete for contracts fuels corrupt tendencies.[3]

Profitability and Competitiveness of the Private Agricultural Extension-Service Provision Business

There is a range of potential service providers that can be contracted by governments to offer extension services. In an analysis of global lessons with contract extension, Rivera and Zijp (2002) report successful public funding/private delivery of extension service models in Chile, where services tended to be provided by private-for-profit firms, in Hungary by universities, and in Venezuela by NGOs. However, where a government chooses to contract for-profit private service providers, as in Uganda, due consideration of business principles is necessary since the same generic principles apply equally to private agricultural extension service provider firms.

For a business to be sustainable, it must have a profit margin that is adequate to meet its overheads, among other costs. The more profitable a business is, the more investors it will attract and the more competitive it will become. With increased competition, the sellers will search for and adopt innovations that will give them an edge over their competitors. The

---

3   It should however be noted that, over time, competition for contracts in on the decline as the program expands to other districts such that the available service providers cannot match the increased demand.

innovations arising from competition will ultimately contribute to the improvement of the overall quality of the services, as the sellers will adopt those mechanisms that enable them to sustainably produce and provide the best quality services in the cheapest manner possible, so as to capture and retain markets. Through improved effectiveness and efficiency, competition leads to an overall improvement in the quality of the services offered and general satisfaction of the consumer.

Mubangizi (2006) found that a general concern expressed by the private service providers involved in NAADS and other stakeholders was that the advisory service and farmer institutional development contracts did not make provision for any profit margin for private service providers. Furthermore, engagement in NAADS private service provision is largely considered to be risky, because of the uncertainty of the contracts and the short duration (3 - 6 months, depending on the enterprise) of the contracts.

A cursory review of the NAADS contract cost structure confirms that NAADS does not explicitly provide for a profit margin for the service providers. The contract structure includes the following costs:

1.   The non-reimbursable professional fee ranging between 200, 000 and 700, 000 Uganda shillings per month (about $114-$402 at the current US dollar exchange rate), depending on qualifications and experience, among other factors.

2.   Reimbursable costs, which include:

    • Operational costs for travel, a daily allowance, stationery etc.

    • Demonstration expenses, for those costs incurred by the service providers in establishing demonstrations.

    • General over-heads for miscellaneous costs in respect of office supplies, postal and telephone costs etc. For these expenses 5% of the service fee is included in the total contract cost.

Service providers operating as firms are hardest hit by this arrangement since they incur other costs not directly related to NAADS activities but which are critical for the existence of the firm in a competitive market. In order to make ends meet, some of the firms have devised means of cutting costs, such as failing to conduct some of the farmer training, paying the technical persons less than what is approved as professional fees, or hiring less qualified staff. Ultimately, this results in less competent and less facilitated people actually delivering the services, with a negative effect on the quality of services.

The perceived unprofitable and risky nature of conducting business with NAADS could, in the long run, adversely affect the quality and capacity of service providers available to render services, due to a failure to attract qualified and experienced professionals, and the inability of those who participate to make sufficient profit to reinvest in their businesses for better capacity.

Adequacy and Sustainability Funding

Agricultural extension service provision is a resource-demanding venture. To facilitate proper planning and implementation required for sustained impact of the services, adequate resources have to be available in a timely manner and on a sustainable basis. NAADS is financed by the government through national, district and sub-county funding, donor support and farmer contributions. The NAADS guideline on funding states that the central government contributes 88% (including 80% from donors), district and sub-county local government contribute 5% each, and farmer groups 2% of the total funds required for the first phase (MAAIF, 2000). It was initially planned contributions by stakeholders would change over the planned 25-year program period, with increasing levels of support district and sub-county government and farmers, and a reduction in the contribution from central government and donors.

Though there is reportedly a steady increase in co-funding by local governments (NAADS, 2005), a number of events place a question mark on the continuity of this trend. The abolition of graduated tax (with effect from 2004/05 financial year) has severely compromised district and sub-county local government's continued and sustained ability to co-fund. In addition, the original assumption of the NAADS design that local governments would generate the co-funding contribution from the expenditure savings with NAADS from reduced recurrent wage costs (MAAIF, 2000) seems to have failed because of the delayed laying off or retrenchment of public extension staff. It was assumed that districts would lay off public extension staff after the introduction of the NAADS program. The savings generated from not having to pay public extension staff would be used to co-fund the private extension system. With these two scenarios, it therefore remains to be seen whether and how district and sub-county local governments will raise their contributions.

A number of studies carried out in different NAADS districts in Uganda provide coordinated evidence that farmers' co-funding cannot be taken for granted. Draa (2005) presents a case study of Arua and Tororo districts, and reports that farmers, especially the poor, reported that they could not afford group membership fees. However, when Draa compared the average wage for a day's work to the annual charges, he concluded that farmers could afford to pay but were either still skeptical about the program, dissatisfied with service delivery or were unappreciative of the value of the advice or information. Draa (2005) further notes that farmers' skepticism about NAADS was further enhanced by NGOs which, unlike NAADS, were providing agricultural inputs and advisory services free of charge. Boettcher *et al.* (1999) are in agreement with Draa, noting that, for farmers to pay for advisory services, they must be convinced that they cannot get the same service elsewhere. Rivera *et al.* (2000) notes that attitude change in this regard among farmers takes time, because of their past experiences with public extension services which did not foster a spirit of valuing technical services. The heavy dependence on external donor support, coupled with erratic contributions from the local sources, paints an uncertain future for the approach.

Clarity of Institutional Roles

The private contract farmer-led approach should ideally involve a range of actors and stakeholders, each with clear roles and responsibilities. The changed roles of government must be spelled out and clearly understood, particularly when a country is changing from a totally public extension system to a system involving the private sector. In many cases, the government's role changes from that of implementing agency to that of providing training and technical information to extension advisors, contract oversight, quality control, program monitoring and evaluation, and overall strategy formulation (Rivera and Zijp, 2002). In Uganda, all these roles are spearheaded by a national secretariat, save for training, which has not been provided for. However, a key issue pertains to the relationship between the private system and the old local government structures.

The new private-sector system hires coordinators at district level, while local governments second technical staff to act as coordinators at the sub-county level without any increment in salary. The agricultural technical staff at district and sub-county level have, over time, assumed the role of supervising the private service providers by serving on the technical audit team. Essentially, the local government structure, consisting of the district production coordinator below whom are technical staff at district and sub-county level, operates parallel to the NAADS structure. Staff assigned NAADS roles are therefore accountable to both the NAADS program coordinator and the district production coordinator. This parallel structure has led to duplication of efforts and, at times, conflict between private service providers, NAADS coordinators and local government staff, perhaps due to the relatively better reward system of NAADS compared to the local government. Private service providers report instances of hostility between themselves and public local government staff who perceive the former as having taken over their work and having better access to resources. The conflict undermines working relationships and sharing of information while the dual roles of local government staff lead to work overload, inefficiency and stress. This could partly explain the deficiencies of the quality-assurance procedures pointed out above. Clarification of roles, reorientation of staff and harmonization of structures and staff reward systems prior to introducing the new contract system could have eliminated these problems.

In addition to government, other actors that should be brought on board include research and education institutions, for their role in providing technical information and training of service providers, the private sector, for provision of input and output markets, agro-processing and value addition and advisory services and farmers. According to Ojha and Morin (2001), for effective cooperation to be attained, institutional roles should not only be clear but also respected by every member of the partnership. Strategies such as memoranda of understanding as well as having an effective coordination body contribute towards the smooth running of partnerships.

The Ugandan system has, so far, established promising partnerships with private companies and organizations for marketing a range of farmers' produce. However, despite budgetary constraints which challenge the partnerships, limited farmer capital, which prevents effective exploitation of the benefits presented by the partnerships, and limited profits, which discourage the profit-minded private sector (NAADS, 2004), these efforts to establish partnerships with the private sector are steps in the right direction.

Conclusions

Information about the utilization of the privatized contract farmer-led agricultural extension approach by Uganda's NAADS Program and elsewhere in the world identify key requirements for success. Among these are a conducive policy environment, appropriate demand articulation procedures and the capacity of farmers to articulate their demands, sufficient private service-provision capacity, efficient and effective service quality-assurance mechanisms, adequate and sustainable funding, conditions conducive to profitable private service provision, and effective coordination of the multi-actor processes that are part and parcel of this complex approach. The following key lessons are emerging.

1. Policy Framework for Agricultural Extension

For farmers to reap optimal benefits from technology and information, government policies that facilitate their access to the full package of other necessary materials and services, including input and output markets as well as credit are necessary. Extension policy reforms therefore need to be accompanied by reforms in other relevant sectors..

2. Developing Farmers' Capacity to Articulate their Demands

Support for farmer institution development processes which are aimed at developing cohesive, representative, sustainable farmer institutions that can serve as a channel for articulation of farmer demand, is a vital component of private extension farmer-led approaches, in contexts where such institutions are either weak or non-existent. This should be complemented by demand assessment processes developed with farmer participation, in an effort to simplify them to suit farmers' literacy and technical competence levels.

3. Agricultural Extension-service Quality Assurance

Mechanisms for ensuring adherence to minimum advisory-service quality standards are a crucial element of successful contract extension systems. Quality assurance needs to consider the quality of the providers, the technology itself and delivery processes. Establishing certified providers of the various services, such as breeders and stock outlets, would make the process much easier. In addition, decentralized service-quality monitoring and evaluation procedures, with participation of farmers' representatives, and input from all target farmers and technically competent personnel is desirable. However, farmers may need to be sensitized about their roles in service quality assurance and their capacities developed to enable them perform this role competently.

4. Private Service-provision Capacity

An adequate and well-functioning private agricultural-extension service-provision sector is an essential prerequisite for effective responses to farmers' advisory service demands. The public sector should be responsible for developing a comprehensive private service-providers' capacity-building and empowerment strategy, aimed at developing their knowledge and skill base, as well as providing conducive terms and conditions for their sustainable operation. The design and implementation of this strategy could require collaboration and coordination between the extension agency and various public sector institutions (such as research and education). The extension agency also needs to make provision for linking private service providers with other stakeholders, notably research and other agencies involved in technical information processing and packaging.

5. Funding

Agricultural extension-service provision is a long-term resource-demanding venture. To facilitate proper planning and implementation for sustained impact of the services, adequate resources have to be availed in a timely manner and on a sustainable basis. This requires adequate and consistent government budgetary provisions as opposed to overdependence on donor support, which is naturally subject to fluctuations due to changing donor priorities.

6. Clarity of Institutional Roles

The private contract farmer-led approach ideally involves a range of actors and stakeholders, each with clear roles and responsibilities as well as terms for executing the respective roles. The changed role of government needs to be articulated and clearly understood, particularly where a country is changing from a public to a private extension system. Clarification of roles, reorientation of staff, and harmonization of structures and staff reward systems prior to introducing the new contract system is necessary to guard against conflict between the old and new structure.

References

Anderson, R.J. and Feder, G. (2003): "Rural Extension Services: Agriculture and Rural Development Department" in *World Bank Policy Research Working Paper 2976,* February 2003. Washington, DC: World Bank.

Bennett, C.F. (1996): "Rationale for Public Funding of Agricultural Extension Programs". *Journal of Agricultural and Food Information,* Vol. 3, p.p. 3–25.

Boettcher, D., Collenberg, D., Huppert, W., Keller, P., Mack, R., Pilgram, K., Schutz, P., Springer-Heinze, and Weiskopf. B. (1999): "Privatization and Commercialisation, a New Panacea? Services for Rural Development". In *Newsletter of the Emerging Platform for Services with Division "Rural Development"* (45) Number 3, December 1999, GTZ.

Chapman, R. and Tripp, R. (2004): "Changing Incentives for Agricultural Extension - A Review of Privatized Extension in Practice" in *ODI-Agricultural Research & Extension Network (AgREN).* Network Paper No. 132, July 2003.

Garforth, C., Angell, B, Archer, J. and Green, K. (2003): "Improving Farmers' Access to Advice on Land Management: Lessons From Case Studies in Developed Countries". ODI-Agricultural Research & Extension Network (AgREN). *Network Paper No. 125,* Draa (200), January 2003.

Garforth, C. and Lawrence, A. (1997): "Supporting Sustainable Agriculture through Extension. Natural Resource Perspective". *Overseas Development Institute, No.21*.

Hall, A.J., Yoganand. B., Rasheed, S.V.and Clark, N.G (eds.) (2001): "Sharing Perspectives on Public-private Sector Interaction" in *Proceedings of a Workshop*, 10 April 2001 ICRISAT: Patancheru, India.

Hoffman, V., (2002): "Farmer to farmer extension – Opportunities and Constraints of Reaching Poor Farmers in Southern Malawi". Stuttgart: University of Hohenheim. 23/07/2003 at 12.35 PM.

Jones, G.E. and Garforth .C. "The History, Development and Future of Agricultural Extension" in B.E. Swanson (1997): R.P. Bentz and A.J. Sofranko (1997): *Improving Agricultural Extension: A Reference Manual*. Rome: Food and Agricultural Organization

MAAIF (Ministry of Agriculture, Animal Industry and Fisheries) (2000): *National Agricultural Advisory Services Program (NAADS)*. Master Document of the NAADS Task Force and Joint Donor Groups 20 October 2000. Entebbe, Uganda.

Mubangizi, N. (2006): Needs and Client Responsiveness of Private Agricultural Extension Service Providers Under the NAADS System: A Case Study of Arua and Tororo Districts in Uganda. Unpublished M.Sc. Thesis. Kampala: Makerere University.

NAADS (National Agricultural Advisory Services) (2005): *Ministry of Agriculture, Animal Industry and Fisheries: Proceedings of the Mid-term Review of the National Agricultural Advisory Services* 31 May to 2 June 2005 at Hotel Africana, Kampala, Uganda.

Obaa, B.J. Mutimba and Semana, A.R. (2005): "Prioritizing Farmers' Extension Needs in a Publicly Funded Contract System of Extension: A Case Study of Mukono District, Uganda" in Agricultural Research & Extension Network (AgREN). Network Paper No. 147 July 2005.

Ojha, G.P. and Morin, R.S. (2001): Partnership in Agricultural Extension: Lessons From Chitwan (Nepal). ODI- Agricultural Research & Extension Network (AgREN). Network Paper No. 114, July 2001.

Rivera, W.M. (1991): "Agricultural Extension Worldwide: A Critical Turning Point in Agricultural Extension" in W.M. Rivera and D.J. Gustafson (eds): *Worldwide Institutional Evolution and Forces for Change*. Amsterdam: Elsevier

Rivera, W.M. and Cary, J.W. (1997): "Privatizing Agricultural Extension" in B.E. Swanson, R.P. Bentz and A.J. Sofranko (eds.). *Improving Agricultural Extension: A Reference Manual.* Rome: FAO Rome.

Rivera, M.W. Zijp, W. and Alex, G. (2000): "Contracting For Extension: Review of Emerging Practices". AKIS Good Practice Note, Agricultural Knowledge Information System (AKIS) Therapic Group: The World Bank.

Rivera, W.M and Zijp, W. (eds.) (2002): "Contracting for Agricultural Extension: International Case Studies and Emerging Practices". New York: CABI Publishing.

Schwartz, L. (1994): "The role of the Private Sector in Agricultural Extension: Economic Analysis" in Network paper 48, July 1994. Overseas Development Administration. London U.K

CHAPTER 3

Challenges and Lessons in Participatory Monitoring and Evaluation:
Experiences of Uganda's Privatised Farmer-led Extension System

Francis Byekwaso, Vincent Kayanja,
Allan Agaba and Grace Kazigati

Introduction

The implementation of a privatised farmer-led extension system by Uganda's National Agricultural Advisory Services (NAADS) is guided by a set of principles including the need to empower farmers to demand relevant advisory services, better targeting of the services, fostering participation and increasing institutional efficiency (MAAIF, 2000). Essentially, these principles constitute key ingredients of a demand-driven service-delivery system envisaged under NAADS. The realization of such a system not only requires articulate and informed beneficiary groups, but also their participation in processes that are critical in ensuring effectiveness of services, including responsiveness to beneficiary needs. Such processes, according to Hilhorst and Guijt (2006), only become meaningful when the primary stakeholders (farmers) are indeed in a position to set goals, track progress, learn from experiences and propose corrective actions for identified gaps. Furthermore, Hilhorst and Guijt (2006) note that, whereas stakeholders' voices could be involved in the planning phase of some interventions, representation of such voices in "conventional" monitoring and evaluation is usually either inadequate or lacking.

A participatory monitoring and evaluation (PME) system that is managed and led by the end users is a vital component of an effective demand-driven research and extension system. In the first place, PME builds the capacity of resource-poor farmers (the demand side) to articulate their objectives for extension services and to evaluate the relevance of the services to their needs. Secondly, the PME approach strengthens local capacity to manage their own projects and take corrective measures where necessary, thereby enhancing the impact of agricultural research and extension products on peoples' well-being. Thirdly, the PME also strengthens producers' capacity to monitor how service providers are meeting their needs.

Essentially, the NAADS PME approach focuses on building capacity to innovate among poor producers. It enables community-based organisations, especially farmer groups, to analyze and interpret progress, to learn from their experiences, and to adjust strategies accordingly. Involving local communities in the PME process strengthens their capacity to demand services effectively, ensures that community perspectives are integrated into development processes and makes these institutions more relevant and responsive to community priorities.

PME, is therefore intended to enable farmers (the primary beneficiaries) become actively involved in monitoring and evaluating program implementation at different levels while, at the same time, providing a basis for systematic self-reflection on what they (the farmers) have achieved in the previous period against what was planned, documented and learnt.

NAADS PME design

The PME was derived from the monitoring and evaluation framework of the program, which provided for incorporation of the disparate voices of farmers into the rural development processes under the program. The PME framework evolved over time, starting with a pilot phase in one sub-county, Kasawo in Mukono District, in 2003, with technical assistance by external specialists from the World Bank and Makerere University's Department of Agricultural Extension/Education. During the course of extending the methodology to other districts, a range of adaptations were made in response to lessons learnt.

Objectives and Scope

The NAADS PME aims at achieving sustainable and effective poverty reduction by empowering farmers to demand more relevant agricultural advisory services. Its purpose is to ensure that advisory services respond to farmers' demands. More specifically, PME targets the primary stakeholders – the farmers – to enable them take the lead in tracking program activities as well as executing timely corrective measures for identified gaps.

In terms of scope, PME is undertaken at three levels – the farmers' groups, parish, and sub-county levels. These levels, also, constitute part of the farmers' institutional structures[4] under NAADS. Appropriate information is generated at each level to inform farmers, politicians and technical staff on the quality of services and any other pertinent issues.

Roles and Responsibilities in PME Process

Adaptation of the process included the articulation of the roles and responsibilities of key actors, and a more formal institutionalisation of the PME process into farmer groups, parish coordination committees and sub-county farmers' fora. Under this arrangement, information generation is a responsibility of the farmers' groups. Each group selects a member to be trained as the group facilitator charged with PME responsibilities. The parish coordination committees, composed of representatives of different farmers' groups within a parish; make a consolidated report from the groups and provide feedback. The selection of group facilitators follows criteria; including literacy in the local language, trainability, a spirit of volunteerism and acceptability by the community (Najjingo-Mangheni and Bukenya, 2004).

It is important to note that, in the latest edition of PME methodology by the NAADS secretariat, monthly reflection meetings by members of the farmers' groups –rather than the lengthy survey interviews originally used–are the main avenues for generation of information. The information is compiled into a farmer group monitoring and evaluation report following consensus by members of each group on the progress made, gaps identified or suggested areas for improvement.

The monthly and quarterly reflection meetings and PME reports are guided by a simple format, which is used to generate the following information:

- Group activities planned and implemented
- Farmer institutional development activities
- Advisory service provision to farmers
- Technology development sites, including their management and utilisation

---

4   The farmers' institutional setup consists of the farmer groups at village level, parish coordination committees at parish level and farmers' fora at sub-county, district and national level. These structures are explained in the previous chapter.

- Replication of technology development sites
- Quantity and type of produce marketed by groups and individual farmers
- Attendance at monthly meetings by each farmers' group

The aim of the reflection meetings is to enable group members share the lessons farmers learnt as a way of improving the program and their farming practices. The parish coordination committee on their part, undertake to compile all issues and lessons for the entire parish. These reporting arrangements also enlist support of built-in feedback mechanisms at each level, where stakeholders evaluate the issues and provide feedback to the relevant levels in the hierarchy as follows:

- Farmers' group facilitators provide feedback to their group members through monthly group meetings.
- Parish coordination committees and community-based facilitators provide feedback to the farmer groups after aggregating the information from the groups in a parish. After intensified technical guidance, some parish coordination committees have been able to adopt the habit of regular monthly meetings.
- Sub-county NAADS co-ordinator and sub-county farmers' fora give feedback to the parish coordination committees and farmers' groups through quarterly and/or monthly field monitoring.

Benefits of PME

The benefits expected from participatory processes like PME are indisputably abundant but, for practical purposes, this section discusses merely those benefits that were observed during the implementation of the PME approach in the NAADS program.

Facilitating improvement in communication and interaction among stakeholders

Farmer engagement with PME in NAADS activities has not only widened the channels through which concerns over and gaps in service provision can be raised but has also provided them with an opportunity to air their views on the desired corrective actions. Furthermore, through the PME reporting mechanism, much more complex issues are brought to the attention of responsible personnel for action. For instance, as a result of this proactive feedback mechanism, stronger interactions between sub-county farmers' fora and parish coordination committees have emerged.

Strengthening and focusing technical audit and quality assurance in service delivery

Information generated by farmers through PME reports has proved useful in guiding supervision and technical auditing. With the PME approach, reports from the farmers (groups and parish coordination committees) provide a basis from which the district and sub-county technical teams can identify areas that need attention. Various districts have alluded to PME contributions for reducing the workload involved in technical audits and the costs involved. This is as a result of increased involvement of farmers in tracking progress and reporting about the services delivered. Unlike conventional monitoring and evaluation which requires supervisors to visit all sites routinely (training centres and technology development sites) as a way of tracking progress and identifying gaps, this responsibility is vested in the farmers.

Availing more reliable information on program coverage

PME reports by farmers' groups provide insights on the number of farmers in each group who access advisory services (attending training/demonstrations by service providers), farmers' groups' co-funding status, as well as the adoption and impact at beneficiary household levels. In Hoima district, for instance, the sub-county NAADS coordinators observed that PME makes it much easier to know the number of farmers attending training sessions and those meeting their co-funding obligations. Other districts, particularly Busia, Kabarole and Tororo, share a similar experience.

Enhancing empowerment of farmers

Besides information generation, the PME approach serves as an empowering process for farmers in the sense that the concrete information generated - when at the disposal of farmers serves as a basis for approving or rejecting the contracts of service providers that do not meet the expected quality of work. Sub-county farmers' fora are relying increasingly on PME reports, not only for verifying of service provider work before making payments, but also for general decision-making in program implementation. Furthermore, greater involvement in program decision-making has led to an increased sense of ownership of the program amongst farmers.

Improving planning of agricultural advisory services

Information from sub-counties shows that information generated by the PME process is used for planning purposes by both management and farmers. The PME process has identified gaps in the provision of advisory services. For instance, in a number of sub-counties, management teams realised that farmers travel long distances to the training centres. As a result of this realisation, recommendations by farmers' groups and parish coordination committees to increase the number of training centres in each parish were taken up and the necessary adjustments made, including considering streamlining these in the NAADS sub-county work-plans in the future. PME reports also increasingly point out the necessity for service providers to draw up their work-plans in conjunction with farmers' groups at parish level to avoid clashes in scheduled meetings. The outcome has been an improvement in farmers' participation in training and demonstration activities in a number of sub-counties. Whereas, in the past, service providers were not required to present their activity plans to farmers this has changed with the adoption of PME. If farmers are to undertake their monitoring effectively, they have to know details of training, such as intended number of sessions, topics to be covered and venues for training. Through PME reports, the importance of such issues has been identified.

Lessons from PME Implementation

As farmers and other stakeholders embrace the PME approach in the districts, a number of lessons are emerging.

PME development is a slow, evolutionary process requiring continuous adaptation

Experience gained so far indicates that a functional PME develops gradually. When PME was initially introduced few farmers' groups could generate meaningful information, but with sustained guidance and practice, most of them are now able to provide the required information on a monthly and quarterly basis. The main hindrances to fast adoption of PME revolve around the literacy of the farmers, and clarity and simplicity in communication by the facilitators.

PME requires trained and committed staff

The interest, conviction and capacity of the personnel directly involved in program implementation are critical to PME success. The commendable progress on PME registered in districts like Bushenyi, Kabarole and Kamuli (NAADS, 2006) is based on effective mobilization and follow-up. These competencies have been built by in-house, systematic and targeted training and mentoring of technical staff, PME facilitators and farmers, based on the capacity gaps that were identified.

PME has human-resource implications

Undertaking PME brought an additional workload to the existing staff, who were already overburdened. Taking on additional staff was found to produce substantial payoffs. In the districts of Kabarole, Bushenyi and Kabale, the presence of young graduates working as volunteers charged with the monitoring and evaluation responsibility greatly contributed to the PME's success. Nevertheless, since the volunteers were brought on board for the short-term- usually 3 - 6 months renewable contracts – PME did not distort the wage bill of the local governments by appointing additional staff. Moreover, once the required capacities have been established the bulk of responsibility for PME is vested in farmers' institutions.

PME works best when farmers are already empowered

Under the NAADS program, PME is primarily the responsibility of farmers who, on a day-to-day basis, are meant to observe program activities, reflect, report, and take corrective action, often with other actors.

The experiences show that groups that are mature, embrace PME quickly. Such groups have functional executive committees and hold meetings regularly as required. Mature farmers' groups also tend to be involved in savings and credit, collective marketing and joint acquisition of inputs. At this level of development, the farmers in these groups are more concerned with increasing their efficiency and effectiveness and therefore willing to take on innovations. It is, however, important to note that at the moment, such functional groups are few – which partly accounts for the rather low levels of PME reporting in most districts.

In addition to the maturity of farmer groups, the level of empowerment among farmers and use of PME are also reflected by the functionality of parish coordination committees and sub-county farmers' fora. The main lesson is that building strong farmers' groups, parish coordination committees and farmers' fora should be prerequisite for introduction of PME. Where such capacities exist, the key roles of analysing information, responding to identified concerns and decision-making are well executed by such organs.

Continuous capacity building is necessary

Efforts by empowered farmers' institutions to carry out PME processes require support from committed and competent program technical staff. Therefore, ensuring the availability of the requisite capacity, especially among program staff and farmers, contributes to sustaining PME initiatives. Farmers' group facilitators need skills to moderate meetings and record data. On the other hand, technical staff need skills related to facilitating group processes, data entry and analysis and report writing. However, obtaining the requisite capacity to facilitate groups of people with diverse and varying interest effectively requires time and effort. In this respect sub-county NAADS coordinators are the key facilitators of PME processes and experience shows that, where relatively uninterrupted and competent program staffing has prevailed, steady progress, particularly regarding PME for program activities, has been registered. In areas where staff changed regularly and/or their capacity was insufficient, PME stalled. Given the continuous nature of PME, high staff turnover undermines continuity and effectiveness. The presence of a local team that understands program initiatives bridges the gaps that could arise as a result of changing circumstances, such as departure and/or relocation of staff. The challenges of coping with an evolutionary process, within which learning and innovation are inherent, can adequately be addressed by focused capacity building.

PME must be integrated within the overall program structure and monitoring and evaluation arrangements

As Hilhorst & Guijt (2006) observe, PME should not be considered as a stand-alone process, but an integral component of a program's monitoring and evaluation. Indeed, for NAADS, the PME strategy is undertaken as a methodological approach within the program's monitoring and evaluation framework, where primary stakeholders-farmers- are placed at the centre of monitoring program processes and impact. PME is anchored in existing farmer and program institutional set up. In this respect, farmers' groups - as grassroots farmer institutions - undertake generation and documentation of relevant information, which is consolidated by a parish level farmers' institution, the parish coordination committee. Potentially, this institutional setup presents a built-in mechanism for feedback to higher levels.

Challenges faced by PME implementation

Although a commendable leap has been made with the initiation and operationalization of PME in NAADS program processes, the following challenges are yet to be addressed:

Quality of information provided

Some groups and stakeholders view the PME reporting formats as bulky. As a result, the forms are filled only partially, leaving many sections blank. It has also been observed that farmers do not always give reliable information. Understandably, most of the difficult sections concern information that requires record-keeping. For instance, farmers need to keep track of costs of production, from land preparation and planting to harvesting and sale, the number of farmers who visit technology development sites and other matters. Until farmers adopt the culture of keeping systematic farm and group records, such information will be difficult to obtain. In other instances, low literacy levels and inadequate training of farmers in PME are responsible for the poor quality of information. However, improvements in the quality of information generated have been observed in groups which have reported for a longer period of time than new groups.

Another problem relates to deliberate editing and distortion of information at different levels of reporting. For instance, if farmers' groups complain about the parish coordination committees, the latter may sometimes edit reports in order to present a positive image of themselves. Remember that it is the parish coordination committees that collect information from the groups and submit reports to the sub-county farmers' fora. In the same way, if the sub-county NAADS coordinator feels "accused", he/she may edit and limit the reporting to positive aspects of the program. However, with a good quality assurance team that is able to verify PME data at the district and sub-county level, these anomalies can be minimised.

Delays in PME reporting due to irregular group meetings

Experience from the NAADS program shows that farmer groups do not hold meetings regularly and yet this is a key prerequisite for PME to work. There are many reasons why groups fail to meet. Credit and saving schemes and farmers' revolving funds provide incentives for farmers to attend meetings. The implication is that PME has a better chance to succeed in communities where there are such incentives for groups to meet, and, while they are meeting to generate information on a timely and regular basis. Delayed reporting by groups affects the submission of parish coordination committee reports upwards to the sub-county local government. In addition to constraints related to irregular meetings, reporting delays are compounded by high illiteracy levels among farmers, such that it takes them a long time to compile the reports.

Weak feedback mechanisms and utilization of PME information

Although a strong built-in feedback mechanism is an integral part of the PME design, it has been observed that little feedback actually takes place in practice. Limited feedback could, among other reasons, be attributed to a lack of monetary incentives for parish coordination committee members to communicate with groups. On the other hand, it may arise because the leaders of the farmers' groups have not yet been fully empowered to own and control their processes without handouts from government. Another serious drawback which has the potential to discourage farmers from active participation is that limited follow-up action is taken by relevant actors on issues raised by farmers in the PME process.

Multiplicity of local languages

In districts like Tororo, the presence of different small tribes in each sub-county causes extra workload for staff, who have to translate formats into local languages. Some sub-counties have resorted to using English, which does not suit all farmers, bearing in mind low literacy levels among peasant farmers. Related to the above, providing formats to groups every month involves the cost of stationery, which constrains the sub-county budget, and consequently some sub-counties do not provide the reporting formats.

Expectations by farmer facilitators

Expectations of remuneration or allowances by some key players is a key challenge to PME. In some cases, parish coordination committees and community-based facilitators expect to be paid allowances for the work they do, especially when they feel that the work is demanding. Other studies (Khanya-aicdd, 2006) show that community-based workers with unrealistic expectations drop out after some time, leaving the work to those who are willing to volunteer their services, or those for whom the services rendered to the community are ample motivation.

Conclusion

PME provides a great opportunity for community-driven development by ensuring that services provided by private service providers meet acceptable standards. In particular, PME is useful for aligning community-level stakeholders so that they identify and streamline their roles in planning, implementation, monitoring and evaluation. Prior to the introduction of PME, the roles and responsibilities of NAADS stakeholders at parish level were not clearly understood. This problem has been resolved with the use of the PME tools and approaches.

The PME approach provides a practical mechanism for obtaining information on program processes, outputs and impacts from beneficiaries. Analysis of PME reports over time promises to be an important source of data for impact evaluation.

In spite of the above strengths, if the PME approach is to be upscaled in all districts and sub-counties, a continuous process of monitoring, reflection and adjustment of the methodology to different needs and contexts will be required.

References

Guijt, I. Gaventa, J. (1998): Participatory Monitoring and Evaluation: Learning from Change. Brighton, IDS. IDS Policy Briefing 12.

Hilhorst, T. and Guijt, I. (2006): Participatory monitoring and evaluation: A process to support governance and empowerment at local level. Guidance Paper. Amsterdam: Royal Tropical Institute.

Khanya-aicdd (Khanya-African Institute for Community-driven Development) (2006): Evaluation of a Community-Based Worker System in South Africa. Bloemfontein, South Africa.

Mihalic, S. Irwin, K. Fagan, A. Ballard and D. Elliot, D. (2004): Successful Program Implementation: Lessons From Blue Prints. In Juvenile Justice Bulletin, *Washington, DC*: USA Department of Justice, Office of Justice Programs.

MAAIF (Ministry of Agriculture, Animal Industry and Fisheries) (2000): National Agricultural Advisory Services Program: Master Document of NAADS Task Force and Joint Donor Groups Entebbe.

MAAIF (Ministry of Agriculture, Animal Industry and Fisheries) (2000): National Agricultural Advisory Services Program: Master Document of NAADS Task Force and Joint Donor Groups; Monitoring and evaluation logical framework, Annex 7.

Najiingo-Mangheni, M. and Bukenya, C. (2004): Participatory Monitoring and Evaluation for Uganda's National Agricultural Advisory Services Program: Development of the methodology, capacity building and piloting. Final Report. Vol. 1. Kampala: Makerere University.

NAADS (National Agricultural Advisory Services) (2005): Participatory Monitoring and evaluation: A guide Kampala.

NAADS (National Agricultural Advisory Services) (2006): Report of the 5th NAADS Government of Uganda-donor Annual Review Meeting Kampala.

World Bank (2005): Action for Social Advancement. Integrating learning in the monitoring and evaluation of CDD projects in the World Bank: a guidebook. Washington DC: World Bank.

SECTION II:
INNOVATIVE EXTENSION
METHODOLOGIES

CHAPTER 4

Reaching Urban
Communities Through Schools:
A Case of the Orange-fleshed Sweet
Potato Project in Kampala, Uganda

Richard Miiro and Boniface Orum

Introduction

Urban people engage in agriculture as a livelihood strategy, often addressing adversity, poverty and other crises. Up to 40% of food consumed in African cities is grown within the cities (Armar-Klemesu, 2000; Nugent, 2000). Bourque (2000) notes that urban agriculture prevails mostly among the poorest sections of some cities, while in others it is a continuation of a long-term tradition due to the availability of open spaces and green belts (Nugent, 2000). Cairo, Havana and Nairobi are examples of traditional farming cities. Among many reasons, people engage in urban farming to produce food for consumption, to enhance household income, and to survive economic crises when market prices for food are high, given that most of the food consumed in the city has to be bought (Nugent, 2000).Obtaining food from urban agriculture reduces family expenditures on food purchases (Nugent, 2000), making food security the most important goal of urban agriculture. It enables families to cope with increasing poverty and high standards of living within cities. Urban agriculture involves mostly poultry and vegetable products, used by urban consumers to improve the nutritional base of vulnerable families, especially preschool children, as is the case in certain areas of Kampala, Uganda. In these areas, urban farming is positively and significantly associated with higher nutritional status as measured by height for age among children, including a lower proportion of severely malnourished children in cases where a caretaker was engaged in farming (Maxwell, Levin and Csete, 1998).

Urban agriculture also provides employment benefits, particularly for wage laborers, and for people involved in aspects of the commodity value chain, such as sellers of tools, seed, manure and fertilizers, transporters and others in the formal and informal sectors. Urban agriculture also contributes to environmental sustainability in cities through the recycling of domestic waste and use of crop and livestock waste as a soil resource (Nugent, 2000; Bourque, 2000).

However, in most cities, farming has increasingly been considered an illegal activity. In Kampala, the capital city of Uganda, ordinances to support the planning of services to urban agriculture were not established until 2005. Extension and advisory services to urban farmers hardly meet their needs, except for livestock and poultry, for which farmers seek support from veterinary doctors. The inadequate servicing of urban farmers can be attributed to their unique nature, the characteristics of the places they farm and, sometimes, gender issues. Farming in cities is mainly conducted by women, who have often gone without technical services (Musiimenta, 2002). Another factor contributing to inadequate urban agricultural extension services is the dearth of methodologies suited to this unique context.

While the global community, governments and city authorities have recognized the importance of urban agriculture and its role in mitigating poverty and environmental degradation, innovative ways of serving and providing the necessary technologies to urban farmers are yet to be fully realized (Armar-Klemesu, 2000). Providing extension services in cities demands innovative methodologies to enhance farming activities in line with city regulations and to the benefit of those who depend on it. Innovative approaches in current agricultural systems in Sub-Saharan Africa require that farmers are empowered to demand and participate in technological development, as well as in the articulation of extension agendas. Empowerment involves engaging farmers in making their own decisions, based on their vision for the future, and engaging in planning, implemention, evaluation and accountability processes.

This chapter presents a case of an innovative technology dissemination approach used to address a food and nutritional security need in Kampala. The approach used schools as centres for learning and dissemination of technology, farmer-to-farmer extension approaches and an array of other extension methods to reach farmers. The approach is based on the extension principle that using multiple methods which reinforce each other results in greater levels of learning and technology adoption. Using school gardens to transfer practical agricultural knowledge to pupils and students took advantage of prevailing government policy that requires all pupils in primary schools to be taught the subject of agriculture because of its critical role in the country's economy.

Reaching communities through schools

The case to be presented involved the use of schools as outreach centers to farming communities in the urban and peri-urban areas of Kampala. Traditionally, the major link between schools and communities has been limited to activities relating to parent and teachers' associations (PTAs) which focus on enhancing pupils' academic performance, teachers' performance and motivation, funding, building projects and at times, school gardens.

Although school pupils and students have often been the targets of new agricultural information and technologies, transferring the information has tended to be an end in itself. Using school pupils and students as a means of reaching and impacting on their communities and, more specifically, a way of encouraging positive change in homes/households has not been fully understood, nor has its potential been fully exploited. However, the important role of schools and the contribution schools can make to sustainable development and ending illiteracy, hunger and poverty, has long been recognized (FAO, 2004). The schools approach is considered to be one of the most cost-effective information and technology dissemination mechanisms because:

- Schools are the building blocks or foundations of our society and shape the leaders and farmers of tomorrow;
- Many schools still have pieces of land that can be utilized to demonstrate a wide range of technologies to the surrounding communities;
- The school gardens can be used to produce food to supplement school feeding programs, thus improving the nutrition of pupils;

- Schools often act as unifying and neutral venues and have facilities like gardens, classrooms, halls and a school compound, to facilitate community gatherings;

- Schools provide the social contexts in which knowledge, behaviours, attitudes, values and life skills are developed and moulded;

- Schools have a mandate to guide young people towards maturity and act as effective vehicles to reach children during the formative stages and hence have high potential of transforming attitudes, habits and character;

- Schools have qualified personnel who need minimum re-tooling/re-training and orientation to teach and train pupils and /or communities on new or innovative ideas;

- The knowledge acquired by pupils can be passed directly to the family members and to wider communities through social activities like music, dance and drama during school open days, science fairs, and community theatre;

- Schools can serve as a channel for community participation and involvement in sustainable education and natural resource management;

- Schools facilitate the establishment and strengthening of vertical and horizontal school-community linkages among related government ministries and organizations like Ministry of Education and Sports; Ministry of Agriculture, Animal Industries and Fisheries; Ministry of Health; Ministry of Lands and Natural Resource Management; National Environment Management Authority, among others;

- Schools provide opportunities for positive interaction among parents, teachers, pupils, farmers, extensionists and researchers and thus enhance interactive learning, information exchange, networking and collaboration;

- Schools are often established institutions with long-term goals and objectives which are essential in providing continuity and sustainability to initiated programs and activities. (FAO, 2004)

The case of the Orange-fleshed Sweet Potato project[5]

The schools approach was adopted by the orange-fleshed sweet potato (OFSP) project to disseminate two improved orange-fleshed sweet potato varieties (Ejumula and SPK 004 Kakamega) to urban and peri-urban farming communities in Kawempe and Rubaga divisions of Kampala city. The orange color of these sweet potato varieties is an indication of the presence of high levels of beta-caroten, a precursor of Vitamin A. When this type of sweet potato is consumed by humans, the beta-carotene is converted into Vitamin A in the body (Louw, 2001; Kapinga Kapinga, Lemaga, Ewell, Zhang, Tumwegamiire, Agili and Nsumba 2003; WHO, 2007). Lack of Vitamin A in the body or Vitamin A deficiency (VAD) can cause preventable blindness and increases the risk of disease (diarrhoea and measles) and death among children. VAD also causes night blindness and increases the risk of maternal mortality, especially during pregnancy. The prevalence of VAD among children and mothers in Uganda makes the promotion of food-based approaches such as growing the orange-fleshed sweet potato critical in addressing the problem (Louw, 2001; Kapinga *et al.*, 2003).

The rest of the chapter shares the experience of disseminating the two varieties, one of the techniques being the rapid multiplication of vines technique - a system that enhanced the availability of planting materials to the targeted urban farmers. The project followed a multi-institutional and multi-disciplinary approach. Partner institutions included the Department of Agricultural Extension/Education (DAEE) at Makerere University, which assumed the roles of overall leadership and coordination of the project activities, community-based planning, community institution building, provision of advice on teaching methods and development of farmer-friendly training modules. The National Agricultural Research Organization (NARO) provided the first batch of planting materials. Together with the International Potato Centre (CIP) it provided technical support, contributed to training module development, and conducted actual training for the project beneficiaries. Joint Energy and Environment Projects (JEEP), a local NGO, was engaged in technology dissemination and supervision, establishment of school demonstrations and the Rapid Multiplication

5   Funding for the project was obtained from the Maendeleo Agricultural Technology Transfer Fund (MATF) of FARM AFRICA.

Technology sites (RMTs), participatory monitoring and evaluation of project activities, and conducting trainer of trainees workshops. Kampala City Council (KCC) carried out mobilization and sensitization of the communities, and provided advice on policy and incorporation of health issues in the sensitization. Schools provided training sites and teachers to train the pupils and engage in community-based activities.

The combination of extension methods and activities used by the project

This is an overview of the various methods and activities used to promote and disseminate OFSP varieties through schools to the surrounding urban and peri-urban communities. The major activities of the project included project-team planning and review meetings, followed by consultative visits to local authorities at the places and schools where the project was to be instituted. The schools and communities identified during the consultative visits were evaluated for their feasibility to work with the project. To establish the extent of knowledge about and use of orange-fleshed sweet potato among the surrounding communities, a baseline survey was conducted, followed by a stakeholders' workshop, at which the project was launched. The communities and schools were approached, mobilized and sensitized about the project with the help of local leaders and volunteers. To ensure follow-up activities, a trainer of trainers (ToT) workshop was held for school and community-based representatives who liaised with the project team and the schools and the farmers.

Training of the pupils was conducted by the school-based trainer of trainers. The practical approach adopted consisted of utilization of live specimens (such as roots, vines, insects, infested plants) and demonstration plots set up at the schools. The demonstrations belonged to the farmers and the school and were co-managed. Planting materials were provided by the project as starter seed for the purpose of home-based growing and observation as well as raising new planting material. Home follow-up of the farmers' and the pupils' activities was done by the project team and the ToTs, while school-based learning meetings for farmers enabled collective assessment of the fieldwork and demonstrations, in addition to exchanging information on good practices discovered. To facilitate further spread of the technologies, pupils were given vines to take home and encouraged to involve their parents in the activities. Pupils and participating farmers were involved in popularizing the vitamin A-rich sweet potato using music, poetry, dance and drama in open-air community gatherings. A detailed description of the process follows below.

The process of implementing the project

The project started by identifying areas in the city to work in. Two divisions considered by KCC to have the greatest need were selected. The divisions were characterized by high prevalence of malnutrition, poverty and lack of development projects. Officials at the divisional level identified 26 schools that could serve as hosts for the project's activities. Feasibility visits enabled the final identification of ten schools to work with. To involve pupils with disabilities, a school for physically handicapped children was included as a project learning center.

Building confidence and teamwork among the partners

The project started off with team and confidence building among ten partner representatives, an average of two representatives per organization. The team building activity involved small-group introductions (person A introduces B who introduces C who introduces A), and setting of ground rules for the project. Transparency, feedback, and self-evaluation were important team activities. The rules were developed using a participatory process. Each member had an opportunity to facilitate a process and contribute to discussions. This enabled all partners to take on assigned roles willingly. Teamwork was mainstreamed at all levels of the project implementation, including among ToTs, the farmers' learning groups and the pupils, where possible. The values and rules set were exercised throughout the planning and review meetings. Using a meta plan to guide discussions and planning was applied.

Selection of target areas and accessing the community

In extension and community development, community entry is considered a very important aspect of community empowerment, particularly with a view to establishing ownership of project activities and encouraging participation. In this case, the participation of community and school leaders was enlisted. The participatory approach was relayed to other activities of the project, such as the consultative visits, during which local leaders participated in selection of working areas according to their needs. Interactive meetings, which allowed local leaders to understand the project by asking questions and making suggestions for improvement, were the main method at this stage. Local leaders further contributed to the project and gave it their endorsement and support during the stakeholders' workshop. The stakeholders' workshop served to sensitize city-level leaders about the nutritional and food security issues that the project was to engage in, as well as solicit city leaders' endorsement of the project.

This was followed by sensitization and mobilization meetings among the selected communities. The methods used at this stage included letters to invite parents of pupils, while the rest of the community around the schools was invited through their local leaders and the use of roadside notices. Both community members and school pupils were sensitized about the project and the role of Vitamin A-rich orange-fleshed sweet potato. The sensitization was done using brief lectures, reinforced with posters of the crop, calendars, animated and attractive text, and question and answer sessions (Box 1). Farmers were engaged suggesting the way forward. This activity prepared community members and pupils to engage with the project. During the sensitization meetings individuals were selected by the community and the school to serve as trainers of trainers or farmer extensionisits.

Box 1:   *Questions by participants*

- If we produced the ORSP, where would we sell it?
- Where are we getting the planting material?
- How do we make the products?
- Don't the vines need special care-spraying with chemicals?
- I don't eat sweet potatoes, do these OFSP also cause stomach problems?
- We brought some vines from Mbale and they failed to grow here, won't these ones also fail?
- Where do we get the machines to produce these products?
- Tell us whether and how the OFSP are different from our local ones.
- Does the OFSP also produce flour?
- But then don't we need machines to make the flour?

Capacity development for technology dissemination

Preparation and capacity had to be established among the chosen communities and schools before full-scale project activities started. Capacity building included sensitizing the community and schools about the project approach, emphasizing on the role of the orange-fleshed sweet potato in combating Vitamin A deficiency and contributing to food security.

Training of trainers

The ground for promoting and disseminating OFSP was laid through training of trainers (ToT), followed by farmer training and demonstrations. Forty two trainers of trainers – 22 school-based and 20 community-based, were trained on the importance of vitamin A, sweet potato agronomy, soil management, the rapid multiplication technique (RMT), facilitation skills, group management and dynamics, participatory monitoring and evaluation. The ToT involved short lectures to convey new technical information, group discussions and projects, site visits and practical activities with demonstrations. The 42 ToTs (men and women) supported the subsequent community training for the rest of the farmers and for the school pupils. OFSP training manuals in English and the local language Luganda a total of 515 copies were distributed to the ToTs, schools and some farmers. The copies at the school were to be used by farmers as reference materials. The future vision was for the schools to serve as agricultural information resource centres for surrounding communities and schools.

Community- level training

Following the sensitization about the nutritional value of the crop, four major topics were handled, i) establishment and management of rapid vine multiplication plots, ii) the agronomy of the new varieties, including pest and disease management, iii) post-harvest handling and iv) utilization. Vines of the orange-fleshed sweet potato were distributed in time for planting during the first rains of 2004 and to establish RMTs and fields. Schools were the distribution centers for the vines. The amount of vines a farmer took was based on how much land he or she had to plant. The initial vines supplied to the farmers were obtained from the technology multiplication centers of the sweet potato program of NARO. Subsequent vines were obtained from farmers and schools within the project area, who had grown clean and plentiful planting material. The school and community ToTs identified farmers in the communities who had clean vines for sale, and also new farmers or old farmers who either needed more vines or had lost their vines to drought. Both the farmers who had vines for sale and those who needed more vines would meet at the school from where the project would buy from farmers with vines and supply vines to those who not only needed vines but had already prepared fields. This method encouraged farmers to maintain clean planting material for sale, thereby earning income

Figure 4.1   *A pupil making planting material for an RMT plot.*

School-based community owned RMTs and fields were established as learning points for farmers whenever they gathered. Observations about the plots were compared to those at the homes of the farmers. They identified ways of handlings challenges and shared practices with each other. Sometimes the pupils of the school worked on the school-community demonstrations under the guidance of the school-based ToT, at other times pupils cultivated their own fields of RMT or the main crop. A method of discovery learning and sharing among participating farmers regarding what worked and what did not work was observed. Suggestions regarding failures were first sought from among the farmers, but usually solutions, particularly about pests and diseases, were provided by the project team members.

Community learning meetings

The first community meeting was convened after invitations were sent through the school pupils to their parents. The school administration, together with the project implementation team sent invitation letters concerning the project and its activities. Farmers who had no pupils in the schools received the message from other farmers and/or neighbors in the community. The project took advantage of the fact that parents and the community in general usually responded positively to invitations from the schools.

The aims of community learning meetings were basically that farmers learnt from each other and shared experiences on OFSP production and consumption. The meetings, which were synchronized with the crop stage, were facilitated by project members. The methods used during the learning meetings were question-answer sessions; alternatively the farmers were divided into groups of 5-8 farmers who discussed various issues and made presentations to the rest of the group. Both methods built capacity among farmers and at the same time allowed active participation by all the farmers. Discussions during the community learning meetings centered around reviews of the roles of ToTs as expected by the farmers, reviewing the progress of OFSP activities, and lessons learnt. For each school-community, one community learning meeting was held every quarter, with an average attendance of 25 farmers per meeting. Meetings were usually scheduled for the afternoon when farmers would be finished with other activities.

Figure 4.2   *(Left): A farmer discussing a point during the farmer learning meeting, and (right): Farmers listening to a project team member as they stand next to a young RMT plot*

Farmers who did not attend training meetings acquired information from friends and neighbors and often from observations of the ongoing project demonstrations. Beneficiaries shared the proceeds of the project with non-beneficiaries and this attracted new farmers to the project. The most commonly used fora for information exchange, planning, implementation, feedback with the project team, farmers and ToTs was through organized meetings. At individual school level, ToTs conducted monthly meetings with farmers and forwarded their deliberations and recommendations to the project team.

Follow- up supervision and monitoring activities

In addition to the school-community farmer learning meetings, field supervision and monitoring was conducted by both the ToTs and the project team. School-based ToTs visited the pupils who had established OFSP fields at their homes.

A checklist that required information about farmers' knowledge, attitudes and skills was used. During monitoring a number of characteristics, including date of planting, development of the OFSP, drought assessment, time of weeding, other effects on the vines, number of vines cut and sold, diseased vines, pests and time of harvesting were recorded. Farmers were also provided with recording sheets and showed how to enter data. These visits also provided for interactions with farmers and pupils in their homes — something they valued and which reinforced their participation.

Generally, the monitoring exercise revealed that pupils gardens were well weeded; vines were healthy and growing vigorously. Less than 10% of the pupils had problems with diseases and pests. Approximately 5% of the fields were affected by straying domestic animals; this was reported byall schools. Ninety five percent of the pupils' parents became interested in OFSP as a result of their children's participation and most were willing to help their children carry out the necessary home-based OFSP activities in their absence (see Box 2 below). Over 70% of the parents visited indicated that they had gained knowledge about OFSP as a source of Vitamin A and the importance of Vitamin A from their children. They had also learnt how to multiply vines rapidly using the RMT technique practiced by their children at home.

Box 2:   *Testimony of one of the parents of a participating pupil*

"I was initially not interested in the child's work because I did not think there was anything special about the sweet potatoes, but after watching her working so hard to establish her RMT plot and mounds I became curious about OFSP. After harvesting, the yields were high and my curiosity turned into interest after tasting and finding that the roots were very sweet. I am now very interested and I am planning to grow OFSP on a large scale" Mr. Kayongo Joseph, the parent of Nambooze Annet of Lubiri Nabagereka Primary School.

Field tours and farmer- to- farmer exchange visits

A field tour to one of the national research institutes was conducted to reinforce farmers' and pupils' knowledge of the source of OFSP planting materials, as well as to expose them to a couple of successful OFSP commercial and model farmers. Field tours also involved school- to-school and community-to-community exchange visits, aimed at sharing lessons, mutual encouragement, and creating a challenge to emulate good performers. A total of 120 farmers and 60 pupils participated in the field tours and exchange visits.

Farmer- to- farmer technology dissemination

To enhance the spread of the technology in the communities, a new set of farmer support members under the "farmer-to-farmer" banner were recruited to compliment the ToTs. Prior to this, there were about 300 farmers actively engaged with OFSP. Ten active farmers per community around the school were requested to identify ten new farmers whom they would mentor into appreciating the importance of OFSP and its production methods. A total of 100 new farmers were expected from each of the 10 school-community centers except for the school for the physically handicapped pupils. About 1, 000 new farmers were expected from this effort. This farmer-to-farmer drive was another activity that was to enhance the knowledge and skills of farmers on OFSP as well as contribute to the dissemination of the technology to new farmers. An incentive[6] for mentoring was provided to the farmer mentors.

---

6   The incentive was an offer by the project to buy vines from older farmers to be given to the new farmers they mentored. In order for a farmer to sell vines to the project, he/she was required to identify, train and mentor at least 10 new farmers in the growing of OFSP varieties. The list of farmers mentored and given vines would then be verified by the ToTs through field visits after which the mentor farmer would be paid for the vine she/she had supplied.

After the exercise, over 300 new farmers were recruited and together received about 137 bags of planting materials. The new farmers were added to another 160 new recruits who had been contacted prior to the farmer-to-farmer exercise. There were more new recruits in peri-urban areas than in urban ones, probably a reflection of more widespread agricultural activities in the former. Kampala School of the Physically Handicapped recruited a school for the deaf to which vines of OFSP were sent. Most of the planting material distributed was from the older OFSP farmers.

Popularization of OFSP through community theatre (music, dance, poetry and drama)

In order to create widespread awareness of the Vitamin A-rich OFSP, a popularization campaign was arranged. The campaign was expected to serve as a promotion and marketing strategy for OFSP planting materials and products (including vines, tubers and processed foods from OFSP). The goal of the campaign was to create awareness about the Vitamin A potential of the crop, its role in human nutrition, food security and household income. The campaigns targeted farming households in the urban and peri-urban areas of Kampala, consumers, children, community leaders and health personnel. The major highlights of open campaigns were music, dance and drama items based on OFSP messages, speeches from various stakeholders as well as exhibits of OFSP products and vines, demonstration of the Rapid Multiplication Technique, distribution of promotional materials and of OFSP vines.

With the help of an independent music, dance and drama trainer, all 11 schools (pupils and teachers) and communities (farmers) were engaged in developing presentation activities to promote OFSP. The presentation included songs, dances, poems and drama often embedded with messages related to the importance of OFSP as a source of Vitamin A, food security, the approach of the project, partnership, poverty eradication, rapid multiplication; varieties of OFSP (Kakamega and Ejumula); soil types favourable for OFSP cultivation, demand, storage, harvesting, the different processed products and project achievements thus far. Music teachers and some talented community representatives from each of the schools and communities were trained for the performance. The OFSP team worked with the lead trainer to develop production and nutritional messages to incorporate into the presentations. Participants in the drama items were community members, school children and a professional theatre group. ToTs trained pupils in their respective schools on OFSP-based drama items while the community ToTs did the same for the community members.

The campaigns were held in open spaces, often near markets or main roads, so as to attract many passers by. Promotional materials for the campaign included T-shirts (over 350), public awareness posters (3000), banners (2) and video documentaries (12).

Achievements and outcomes

The major achievement of the project was the successful implementation of an innovative approach in which multiple partners and multiple extension methods and schools were combined to produce a multifaceted effect. As a result, there was widespread adoption of the Vitamin A-rich orange-fleshed sweet potato in the project areas and beyond. Even though the project met a number of constraints, such as timeliness of activities, and drought, out of an estimated 6,000 bags of planting material distributed (equivalent to about 300 acres of OFSP), 60 tons of tubers were harvested. This was short of a potential yield of 1, 800 tons which would have been obtained from 300 acres if all conditions had been optimal. In addition to the tangible outcomes, intangible benefits included increased visibility and respect of schools that were involved (see Box 3) and stimulation of markets for vines and tubers.

The experiences of farmers and pupils on a wide range of skills ranging from people skills such as communication and group management to technical skills like raising RMTs, producing clean materials, and making processed products out of the OFSP resulted in the development of a cadre of local resources, people who can be tapped by other development initiatives.

Box 3:   *Testimony from the headmaster of a participating school*

According to Mr. Charles Senkungu, the headmaster of Cleveland and Hill Day and Boarding Primary School in Kawempe, his school achieved the following: the teachers and community members were trained as ToTs, the school hosted and worked with neighboring communities to grow, multiply and promote OFSP; 100 to 200 school pupils benefited from the OFSP; the school became known because of OFSP and recruited more pupils.

Challenges of implementing the project

The challenges met during the implementation of the approach can be categorized as contextual/environmental, technical and organizational/ institutional. The contextual and environmental challenges stemmed from the urban and peri-urban nature of the areas and natural factors. These included drought, which affected yields and maturity of the crop and the limited land sizes which meant that the RMTs had to compete with other crops. Due to the limited land, some farmers took the materials to villages where more land was available. Despite the innovative approach in which different methods were used to encourage technology spread, the limitation of land sizes permitted little to be done. For example, too little was produced to enable sale of tubers and surplus for processing. The demand for OFSP could not be met. Nevertheless innovation continued, with some farmers finding niches in handling processed OFSP products.

The technical challenges were related to the technology itself. One of the varieties (Ejumula), despite its high yielding quality and processing ability was prone to diseases, particularly the sweet potato virus. The other variety (SPK 004), despite its slight disease resistance, was prone to drought. These challenges often affected the morale of farmers. However the constants presence of and encouragement by the project team and ToTs maintained optimism among the farmers. The rapid multiplication technique (RMT) for multiplying vines received low adoption despite being appreciated for its role partly due to the more convenient traditional practice of obtaining vines from old fields, the lack of land for establishing RMT plots, and the need to irrigate since RMT plots have to be planted before the planting season.

Organizational or institutional challenges related to the nature of the project within which the approach was implemented and the multiplicity of institutions involved. Being a project, activities were often interrupted when main institutional work took precedence. Handling the number of schools and communities and ensuring adequate follow-up of the changes involved hard work for the project team. Working with the trainers of trainers who were largely volunteers presented a challenge since not all were consistent. A proper reward system was necessary. Even for the most committed ToTs, the work of following up tens of farmers as well as caring for their own work posed a challenge. Thus coordinating a multi-layer team and all the activities needed a lot of time, teamwork and good planning.

For the schools, despite the fact that the project component of involving pupils was supported by a government policy of teaching and vocationalizing agriculture, and the willingness of the schools to partner with the project, challenges arose. The most important challenge was the disruptive nature of the activities to school programs, often requiring mainstreaming project activities with school plans. Sometimes conflicts arose between the school leaders and the school-based ToTs or between different teachers struggling to participate in the project and wanting to share.

Lessons Learned
Managing inter-institutional collaboration

Successful technology dissemination through inter-institutional projects involving schools requires agreement on certain principles: a good working relationship amongst project implementers, school administration and farmers; transparent leadership at the school, community and project level; sharing of roles and responsibilities; participatory decision-making, flexibility, mutual respect and trust, constructive criticism, and effective feedback mechanisms. In addition, teamwork which requires agreement on principles of working together, is essential.

The use of a participatory approach at all levels from the identification of project target areas, to building local capacity and implementing the program contributed greatly to the project results. In addition, active participation by local leaders, school head teachers, school-based ToTs and community farmers helped to maintain the spirit of cooperation between the farmers, pupils, ToTs and the project implementation team.

Need for adequate human, financial and other resources

Successful implementation of the approach hinged on the existence of all the necessary expertise (including expertise on the technology itself, as well as extension and participatory methods) within the project team.

The availability of resources to ensure that the project team and its community-based support cadres reached the community, supplied the planting material and conducted supervision contributed greatly to the success of the project.

Use of complementary multiple technology-dissemination and capacity-building methods is more effective

Use of different approaches to community engagement, namely consulting with local leaders at different levels, sensitization seminars, conducting community learning meetings, home and field visits, establishment of community structures, helped to create legitimacy for the project and thereby maintaining stakeholders' interest, enthusiasm and opportunities to provide feedback on the progress of the project.

Closely linked to use of schools as the center of technology dissemination was the ability to reinforce the OFSP key messages through different information dissemination channels, such as community training, seminars/ conferences, exchange visits, school and community theatre, participatory planning and monitoring, radio messages and information transfer by school pupils.

Follow-up, monitoring and evaluation are crucial

Regular documentation and field monitoring by means of a good monitoring framework/protocol is not only crucial for collecting, analyzing and reporting data, but also for making the project implementer' presence felt on the ground.

Conclusions

The use of schools as outreach centres to urban and peri-urban farming communities through a multi-institutional and multi-extension methods approach proved useful for spreading information about the Vitamin A-rich OFSP. It is the multiplicity of methods, the teamwork and the expertise applied to the implementation that produced the results shared above.

Among the key ingredients of the approach were capacity development of community and school-based workers and involving farmers as mentors of fellow farmers. The community-farmer learning groups served as useful fora for farmers to exchange ideas on what worked and did not work and how to handle difficulties. The potential of school pupils as technology disseminators and as influential members of families and communities was demonstrated. Several extension methods, augmented with participatory methods, were used to create awareness, conduct meetings at team and community level, develop linkages, popularize the technology, ensure farmer-to-farmer learning and leadership development and plan for the future.

While the approach was suited to the unique urban context, it can also be adapted for rural settings, as has been demonstrated by one of the partner organisations, JEEP, which applied it in a rural district. Kampala City Council urban agriculture department was also able to adapt aspects of the approach to implement a vegetable growing project in schools around in the city with funding from an NGO, PLAN Uganda.

References

Armar-Klemesu, M. (2000): Urban agriculture and food security, nutrition and health. Thematic Paper Number 4. In Growing Cities Growing Food: Urban Agriculture on the Policy Agenda: A Reader on Urban Agriculture. DSE GTZ CTA SIDA. p.p 99 – 117. Accessed on 19 of February 2007 at www.ruaf.org/node 58

Bourque, M. (2000). Policy options for urban agriculture. Thematic Paper Number 5. In Growing Cities Growing Food: Urban Agriculture on the Policy Agenda: A Reader on Urban Agriculture. DSE GTZ CTA SIDA. Pp 119 – 145. Accessed on 19 February 2007 at www.ruaf.org/node 60

FAO (Food and Agriculture Organization) (2004): School gardens concept note: improving child nutrition and education through the promotion of school garden programs. Special Programme for Food Security (SPFS) handbook series SPFS/Doc/31. Rome: FAO.

Kapinga, R. B., Lemaga, P., Ewell, D., Zhang, S., Tumwegamiire, S., Agili, and Nsumba, J. (2003) Increased promotion and evaluation of high ⏥ carotene sweet potato as part of the food-based approaches to combat Vitamin A deficiency in sub-Saharan Africa (SSA). Accessed on 10 March 2007 at http://www.cipotato.org/vitaa/Publications/FOOD%20AFRICA-OFSP%20Promotion%20and%20extended%20.paper.doc

Louw, M. (2001): Vitamin A. *The Medicine Journal.* Vol. 43. No. 4. Accessed on 10 March 2007 at www.medpharm.co.za/safp/2001/may/vita2.html.

Maxwell, D., Levin, C. and Csete, J. (1998): Does urban agriculture help to prevent malnutrition? Evidence from Kampala. FCND Discussion Paper 45 Accessed on 20 February 2007 at http://www.ifpri.org/divs/fcnd/dp/papers/dp45.pdf

Musiimenta, P.T. (2002): Urban Agriculture And Women's Socio-Economic Empowerment: A Case Study of Kiswa and Luwafu Areas In Kampala City. Accessed on 20 February 2007 at http://www3.telus.net/public/a6a47567/KampalaWomen.doc

Nugent, R. (2000): The impact of urban agriculture on the household and local economies. Thematic Paper Number 3. In Growing Cities Growing Food: Urban Agriculture on the Policy Agenda: A Reader on Urban Agriculture. DSE GTZ CTA SIDA. Pp 67 – 97. Accessed on 19 February 2007 at www.ruaf.org/node 57

WHO (2007) Micronutrient deficiencies. Accessed on 10th March 2007 at http://www.who.int/nutrition/topics/vad/en/

CHAPTER 5

A Transformative Approach to Extension:
Innovative Technology Transfer Methodologies for Plantain (Matooke) on the Ecuadorian Coast[7]

Colette Harris and Carmen Suárez

Introduction

Historically, a serious issue for the development of agriculture has been the gap between farmers and researchers, especially for resource-poor farmers in the global south. This gap has been attributed in the first place to a failure on the part of scientists to engage with farmers at all stages of the research process, from identifying the question to be addressed to collaborative development of appropriate technology-transfer methodologies, something only partially mitigated by the use of participatory approaches to planning and evaluation (Werner, 1993: 13).

This is because of the complex nature of the reasons for failure to adopt, which may not be fully understood by scientists or extension workers. For instance, in Ecuador, discussions with farmers exposed the fact that they felt uncomfortable about changing the way plants have always grown. Cacao farmers saw witches broom tubers as integral parts of the plants rather than as pests, while plantain farmers thought that all new offshoots should be given the chance to survive and were therefore reluctant to thin them out.

Another reason for failure to adopt new methods is that farmers may fear they are too risky. The poorest farmers operate on a tiny margin between survival and bankruptcy and they cannot afford to jeopardize their livelihoods on something not completely tried and tested (Scoones et al, 1996: 9). It follows therefore that it is the richest farmers, those who can best afford the extra time and the risks involved in experimentation, who are most able to collaborate on on-farm research projects.

---

7   The project of which this was a component was carried out under the auspices of the USAID-funded Integrated Pest Management Collaborative Research Support Program (IPM CRSP) managed by the Office of International Research, Education and Development at Virginia Tech.

In many countries, the group that has the least exposure to new technologies and extension services is women. Not only do female farmers tend to be among the poorest but they are seldom, conceptualized as decision-makers on family farms. They may thus be considered ineligible to participate in research projects and their viewpoints may not be sought.

Even when extension services reach farmers, this is often by way of top-down, simplistic approaches, such as lectures, with occasional field days to demonstrate certain techniques. Extension service providers show farmers how to apply techniques but fail to take farmers' responses or ideas into account (see Ramirez, n.d. :6). Moreover, even when they do use participatory methodologies, extension service providers are seldom trained to facilitate in such a way that farmers develop their own learning skills. Instead, they tend to explain what to do and simply repeat it until the farmers understand it well enough to use it for themselves, rather than helping farmers grasp the principles behind it. The problem with this is that if the agro-ecosystem environment changes even a little, the technologies may fail, leaving farmers in disastrous circumstances. In other words, the use of participatory methodologies has failed to make significant improvements in the adoption of new technologies (Nathaniels, 2005: 3).

The Farmers' Field School Approach

Starting in 1990, Indonesia was the cradle of a new and exciting extension methodology-the farmers' field school (FFS). This went beyond the merely participatory to develop an approach whereby farmers learned how to make their own diagnoses and conduct on-farm research. At the start, professional extension workers were used, but later the involvement of an increasing number of farmers allowed new schools to be facilitated by farmer trainers (Schmidt, 1997).

In this approach, farmers and facilitators come together to decide on a learning agenda and then work together on a weekly/fortnightly basis on the crops throughout the growing season, with sessions for perennial crops being organized around phonological cycles. The schools are often organized for the purpose of teaching integrated pest management or occasionally integrated crop management.[8] One of the most important

---

8   These are two conceptually different approaches-the former working to minimize deviations from a supposedly correct growing cycle, the latter dealing with crops holistically.

features of the schools is that they are learner centred. Farmers are not lectured at but shown ways to explore the workings of their local agro-ecosystems. They are encouraged to make their own observations in regard to all aspects of this. In this way, farmers learn to understand causes and effects of agricultural processes that influence the crops they are growing, which allows them to make informed decisions.

This was the approach taken by the original Indonesian schools (IPM, 1996). One idea behind this kind of educational process was that, beyond learning how to deal with current problems, if farmers gain a real understanding of the principles on which agro-ecosystems function they will be able to adapt to new circumstances and to evaluate new information that comes their way relatively easily. However, unfortunately, only too often this ideal is not realised because many extension staff are unable to support farmers to learn for themselves, so that participants in FFSs learn very little more about how the environment functions than they would have from lectures. It takes a real commitment to a different way of working and ongoing training of the extension staff to produce significant change.

The second issue is that of resources. FFSs require not only a large time commitment on the part of both extension staff and farmers, but are also costly in terms of other material resources supplied for the training. Consequently the number of farmers reached, even over a number of years, tends to be relatively small, especially given the extent of expenditure. A further problem is that some extension staff give farmers so many tasks to complete between their weekly sessions that they cannot manage these in addition to their other work. This produces frustration and encourages farmers to drop out.

The costs of the FFS are, to a certain extent, offset by the fact that participating farmers are encouraged to pass on what they have learned to others in both formal and informal contact. The latter may occur with any group of farmers with whom the FFS member has contact or even by other farmers observing FFS members' fields.

Studying FFS carried out on green beans under the auspices of the IPM CRSP in Mali, it was found that incorporating women into the schools made a significant difference to their impact on the community at large. Once the women joined in, most of the village became aware of the schools and a high percentage of community members applied at least one of the technologies that were being taught, while previously this knowledge had been limited largely to male participants (Sissoko, Sidibe, Harris and Moore, 2003).

It should be noted, however, that the Nathaniels report suggests that the information passed to the rest of the community by FFS members consisted largely of simple messages around particular technologies or seed varieties and tended to exclude those aspects of the training that taught reasoning processes and more complex ideas such as insect lifecycles (Nathaniels, 2005: 6). The Mali study referred to above reached the same conclusions (Sissoko *et al.*, 2003).

Formal transfer of information occurs when FFS members become facilitators in their turn and start their own FFS. In theory, this should mean that new generations of participants gain insights similar to those of the first generation. However, this necessitates a high level of capacity for this kind of work and good training of the farmer facilitators. Obviously, not all farmers are suited to this but, if the first generations of farmers are carefully chosen for their teaching abilities, a high proportion may be capable of becoming facilitators.

One female FFS trainer in Benin suggested the following characteristics as important for this task-knowledge of the FFS curriculum, availability in the village, an easy-going nature, and the ability to read and write' (Nathaniels, 2005: 5). This is somewhat alarming because it suggests that there was a set curriculum and that new approaches to conceptualization were not encouraged. The authors believe that good facilitators need an ability to allow participants to think things through for themselves and to ask appropriate questions to stimulate thought. A certain level of analytical reasoning is just as important, as is sufficient knowledge to be able to cope with new on-farm issues that arise, even if only by passing the information on to researchers. However, to learn to do this is not easy, especially if the original extension staff did not operate in this way, so that the farmer facilitators have no role models. Finally, the FFS also needs logistical and other support, including a high level of resources.

Another issue that needs careful planning is the incorporation of women. In some settings women will not be allowed to participate in mixed groups, in which case it will be necessary to have separate schools for them. In others, while cultural barriers do not prevent mixing, time constraints may make it difficult for women to participate along with the men, since they may have to take time off in the middle of the day to perform domestic tasks, or else women are so unused to speaking in front of men that they remain in the background and fail to develop their own capacities. In some settings, including much of Ecuador, it is difficult to find out how much women actually participate in agriculture and thus, how much they would benefit from incorporation. In sub-Saharan Africa, however, women make up the majority of farmers, while still being very much the minority of extension workers as well as of those reached by extension services.

Outside the field sessions, many FFS facilitators also provide technological information. If they do this in lecture form, as a number of the Malian FFS trainers did, this may prove difficult for illiterate farmers to follow. This is again very often a gender issue. In Ecuador it is the men who tend to be less educated, while in Mali, as more generally in Africa, it is the women. The Malian FFS study showed that illiterate female farmers did not learn well when presented with theoretical concepts by way of semi-lectures, while the more educated men found this a good way of learning, especially those sufficiently literate to take written notes. Indeed, there was a noticeable discrepancy in knowledge acquisition between those functionally literate and the rest, irrespective of sex (Sissoko, *et al.*, 2003). This was compounded by the fact that the facilitators did not use a teaching process that encouraged the participants to reason through the technological information. Instead it was presented as a given and the participants are simply expected to learn it by heart.

The teaching methods used, as well as other factors concerning willingness to adopt new concepts, can be linked to a phenomenon found in both the Benin and Mali studies – the fact that farmers tend to be selective about which processes they found interesting as well as which they passed on. The Benin study found that farmers rejected some technologies outright and were indifferent to others.

The conclusion was that the farming practices on offer in the FFS did not always meet local requirements. Reasons given for not adopting certain practices included peer pressure to continue using agrochemicals rather than adopting the integrated pest management or organic methods taught in the schools, the inability to afford all the materials necessary, lack of access to labour or inputs, terrain-related difficulties, and fear of lower yields (Nathaniels, 2005: 6-8).

Experiences in conducting schools in Ecuador have shown that the reasons articulated for lack of interest often conceal fear or shame on the part of farmers, or even extension workers, for not having understood the practices or the reasons behind them. For instance, despite plantain farmers having been told on numerous occasions that they needed to remove the old plantain plants completely after harvesting, they took no notice until after studying the life cycle of the banana stem borer or black weevil. Only after understanding fully that the insect lays its eggs in these old plants and thus uses them as breeding grounds for the production of larvae that then enter the adjacent new plants, did the farmers decide not only to clean up their own plantations but to mount a campaign to encourage neighbouring farmers to do the same and thus to reduce the overall population of weevils.

This suggests the benefits of a discovery-based approach and the importance of incorporating farmers' ideas in developing extension instruments. Both the Benin and Mali studies showed that when farmers are involved not only in learning new packages but also in their development, the level of uptake is considerably higher. However, one of the problems with farmer-to-farmer approaches is that this makes it more difficult to convey new ideas and issues to researchers. Incorporating mechanisms for conveying information from farmers to researchers and not only the other way round could provide useful feedback into the research process. However, at present this is rarely done.

As a result, communication between farmers and scientists tends to be reduced largely to scientists endeavouring to convince the farmers to adopt technologies they may neither understand nor want, while that between farmers and extension workers is blocked by the paternalistic and authoritarian manner of the latter. In any case, it is difficult for extension agents to keep up with new scientific developments, especially in the south where they may not be able to use the Internet for this, nor have contact with agricultural scientists, and literature may be couched in complex technical language. The upshot is that farmers are left to fight problems on their own or buy agrochemicals touted by the salespeople who are very often the main providers of agricultural information.

This was very much the situation found in Ecuador in regard to plantain farming and it was in response to this situation and demands by farmers for access to improved information sources, that the program was developed.

A new approach to extension

The initiative to work on developing alternative extension approaches came from a meeting between the two authors of this paper, Carmen Suárez providing the agricultural information and Colette Harris developing the extension methodologies.

Plantain had recently become an important export crop for farmers of the coastal region of Ecuador. Exporters were demanding that plantain farmers use the same agrochemicals that had long been applied to sweet bananas without taking into consideration either the differences between the crops or the harm these chemicals were causing to human health and the environment. Harris and her research team wished to counter this tendency by developing holistic crop management approaches that used a minimum of chemicals. They did not wish to do this in isolation in their laboratories but rather in the field with farmer input and support (Suárez-Capello et al 2001, 2003).

In order to obtain a more detailed understanding of current farming practices and attitudes, a survey of 300 male and female household heads was carried out in 2003 to learn about farmers and their families, as well as about the state of their plantations, their knowledge of crop management, their use of agrochemicals, and their market-related practices. At the heart of the survey was a set of photographs of plantain plants showing the most common symptoms of the pests and diseases most frequently encountered. As a follow-up, a team of three mixed-sex pairs carried out in-depth participant observation over a period of some months with six families, which provided detailed information about their daily lives and farming practices (Harris, Vera, Cabanilla, Cedeño, Barrera, Suquuillo, Crizon, Suarez, Norton and Alwang, 2003).

These two surveys provided Suarez and her team with important information. In the first place, they were able to note the terms used by farmers so they could make use of these terns rather than technical terms in their interactions with farmers. This greatly improved communication. Secondly, the use of photographs allowed farmers to articulate their understanding of the symptoms of pests and diseases seen in their plantations and to name what they believed to be their causes. This gave the scientists a good picture of farmers' conceptualization of their phytosanitary problems and helped them realise why so many of the recommendations given by extension or researchers had not been adopted. This appeared to happen whenever the advice did not fit in with local conceptions of how plants and the agro-ecosystem in general functioned.

While the research team started to work more closely with farmers to develop approaches to plantain farming they hoped would respond to the real needs of the farmers and thus be readily adopted by them, they also began to think about how they might reach the vast number of farmers involved. Statistics showed there were some 5, 000 plantations on the Ecuadorian coast farming plantain commercially (Orellana, Unda and Analuisa, 2002).

It was clear that the existing extension services had only managed to reach a fraction of these farmers and that the methods extension was using, mainly lectures along with the occasional field day, were not succeeding in imparting much information. While, in the highlands, FFS had been far more successful, very few farmers had been reached while the annual cost was a high $300 US per person.[9]

Suárez and her group on the coast wanted to reach far greater numbers, at a much lower per capita cost, and without the constant presence of an extension staff. Harris was charged with the task of developing a methodology to fit this ideal. It would have to be accessible to all educational levels and learning capacities in order to reach the entire group of farmers and to permit them to explore their environment and learn to think through cause and effect without the presence of an extension staff, a method aimed at providing the optimal learning experience for both literate and the illiterate farmers.

The development of the course

The first challenge was to provide a similar level of exploration of the agro-ecosystem through the use of photographs and other materials as that which is provided in FFS through weekly group observations. Moreover, the idea was not to *tell* the farmers what they should know but rather to encourage them to work through the causes and effects for themselves.

---

[9]   As calculated by a Virginia Tech graduate student in agricultural economics, María Manceri.

In order to do this, the researchers needed to come up with methods that would help farmers to perceive relationships in situations where these were not at all easily realized. There are many situations in the field where cause and effect cannot be perceived through simple observation, but need other approaches.

Moreover, there are many processes around plantain plants where cause and effect are anything but obvious. First, there is the structure of the plant itself and the way in which it develops. To the farmers' eyes the hand of the banana appears to develop suddenly at the moment when it appears. In reality, plantain plants grow fruit in very much the same way as babies grow in the human womb. That is to say, the bunch exists inside the plant in miniature form. It starts off low down and as it gradually rises it grows in size until one day it comes into sight, largely fully grown at the top of the plant. Farmers think that it is unnecessary to provide fertiliser prior to this moment, since they see the fruit as only needing nourishment "from birth" while, in fact, just as with a human foetus, good nutrition is vital from conception. In the case of plantain, carefully dissecting plants at different stages of growth allows farmers to "discover" a tiny but identifiable bunch lying well inside the plant and to understand when it is really formed and how long it takes to grow into maturity.

Then there is the issue of pests and diseases. Among the most prevalent on plantain in Ecuador are Sigatoka and weevils. Sigatoka is a leaf disease spread by tiny, almost imperceptible, spores. It is necessary to understand how these travel from leaf to leaf in order to comprehend the importance of limiting infection by cutting out damaged parts as soon as it appears. A further important issue is the fact that the damage produced by the spores is not immediately visible but takes some time to appear. As regards the weevil, it is necessary to understand its lifecycle in order to grasp how it damages the plant. Most farmers are unaware of the connection between the large worm-like objects or larvae they see in the corm and the adult weevils. Some even believe weevils damage the plant by flying up to the fruit and eating it.

For the development of the course it was necessary to find ways of supporting participant farmers, to help them reason through these issues. For instance, as regards weevils, some way of understanding their life cycle was necessary, while for Sigatoka, farmers needed to understand the fact that it is caused by spores even when these remain invisible to the naked eye.

The development of the materials for the plantain course was carried out with the close collaboration of local farmers. A large number of photographs were required, as they would form the basis for the course. Here a digital camera was used, which allowed the course developers not only to judge for themselves on the spot whether the photos were useful but also permitted them to work with the farmers in taking and selecting photos, to ensure that the farmers' vision and not the researchers' prevailed and thus that farmers would be able to relate to these photos appropriately. This task enabled the researchers to discover that farmers were adept at observing many details that scientists tended to miss.

Once there were sufficient photos they were grouped in sets by topic and accompanying questions were developed to focus attention on specific issues. In fact, the course was to consist entirely of open-ended questions, with no statements being made by the facilitators. The intention was to open up spaces for reflection and discussion, while statements tend to close spaces and predetermine ideas and opinions.

The trajectory of the plantain course

The course starts by allowing farmers to identify their problems and set their own timetables. Next then they tackle first the sanitary and later the pest and disease problems of their plantations, one at a time. During the first session farmers are asked to demarcate a demonstration plot in which they can test the practices they learn during the course. They then conceptualize the current state of their plantations and identify the chief problems confronting them by selecting from a wide variety of photographs taken across many different types of plantations, as well as of specific issues relating to pests and diseases. Finally, they establish the order in which they would like to deal with the problems.[10] They do this by pasting on to a cardboard ladder photos of the activities needed to improve their farms and

---

[10]   It should be noted that, to date, the farmers have always chosen as their ideal the demonstration plot farmed in accordance with the principles of integrated crop management by project technicians over the very best of their own plantations. During the trial stages the discussion around this question gave the researchers the opportunity to note down the terms used by the farmers and also to judge the state of their knowledge of the biological processes taking place on their plantations.

the order in which they should be carried out, starting with the present state of their plantations at the bottom and ending with their ideal plantation at the top (Figure 5.1). In between, the farmers place the activities they think should be carried out to arrive at this ideal in the appropriate order.

Figure 5.1 *Ladder of farming activities*

The other modules follow the plan laid out on the ladder and they also use photos and questions, along with, in some cases, observation exercises and field experiments the farmers are asked to work on between sessions.

The module to help the farmers understand the importance of a holistic approach to crop management, with particular emphasis on the plantation's sanitation and on good nutrition the primary two requirements for maintaining a healthy crop – made use of a round styrofoam barrel made of labelled segments to represent all the crop's needs (Figure 5.2).

Figure 5.2   *Styrofoam models for teaching the importance of nutrition and sanitation*

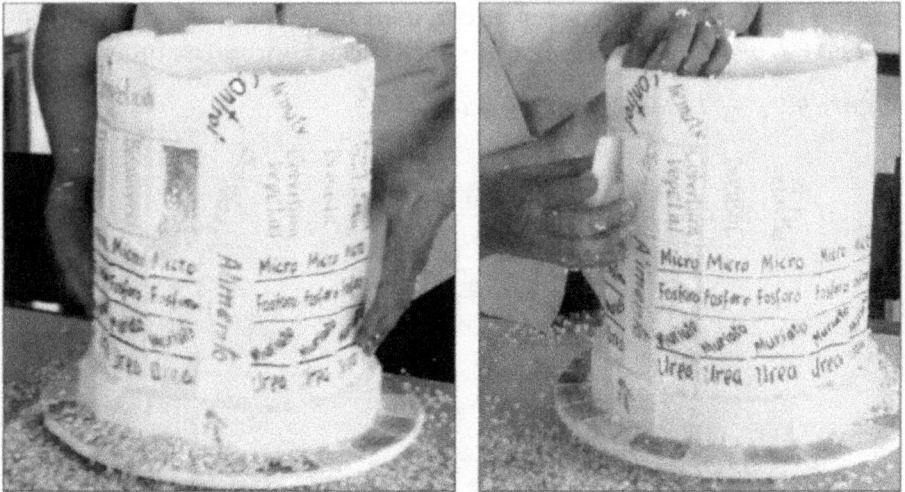

Styrofoam peanuts were placed inside this barrel and this permitted the farmers to see that eliminating any one of these segments allowed a percentage of the "crop goodness" to leak out and thus weaken the plants and made destruction by insects and diseases easier, just as happens when the human body is not well fed and kept clean.

Two of the most important modules are those that deal with the banana stem borer and with black Sigatoka, the chief insect pest and disease of plantain in Ecuador respectively.

In the former, as suggested above, the main issue is to help farmers understand the form in which the weevil damages the plants. Thus, children are encouraged to help their parents by conducting experiments, for instance by capturing the larvae of weevils and related insects and feeding them until they emerge as adults. It was the act of watching the actual process of metamorphosis of the different types of weevil and seeing the adults emerge at full size that each type is found in their plantations, that convinced farmers both of the relationship between larvae and adults and of the fact that there are different types involved. This process helped sharpen the farmers' observations so that they subsequently discovered for themselves that only one type of weevil is found on or near the live corm. All other types in the plantations are found only on plants in the process of decomposition (Figure 5.3).

Figure 5.3   *Different stages of metamorphosis of the banana stem borer or black weevil (Cosmopolites sordidus)*

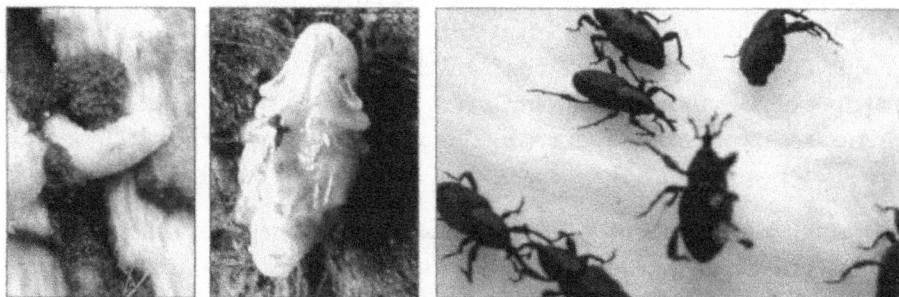

Another insect commonly found on plantain corms is the earwig. Farmers have long considered these to be harmful to the plant. However, by capturing them and exploring what they actually eat, the farmers have realised that in fact the earwigs feed not on plantains but on weevil larvae and thus are beneficial insects. Therefore, when the farmers find them on the corm it is because they are helping the farmer getting rid of larvae not because they are harming the plant.

Once farmers understand the issues discussed above they learn that they can steer banana weevils away from their live plants by placing chopped up pieces of the corm of plants already harvested between the rows of live plants and making access to the corms of the latter more difficult and less pleasant. In the past they had not bothered to cut down the harvested plants since they considered this a waste of time and even thought the older plants provided a source of nutrition for the young ones. Now they find that by merely removing all the old plants completely and using the chopped up corm as an insect trap they can significantly decrease weevil damage.

Thus, by a series of entomological experiments, the farmers were able to learn enough about the relevant insects to be able to figure out how to protect their plants. They now realized that it was better to encourage the insects to lay their eggs elsewhere and help the beneficiaries to flourish than to spray insecticide, which would kill off all insects, including beneficial.

This is also advantageous to human and environmental health, since the chemical commonly used on these insects is one of the most toxic on the market.

In the case of Sigatoka, comparisons with human viruses, such as the flu, were made using graphics. In this way the farmers were brought to realize that, as with human diseases, plant diseases pass through the air invisibly to the naked eye. Strong magnifying lenses allowed them to see the infected patches of Sigatoka on plantain leaves and the growth of the fungus on them. Two experiments that enabled them to perceive key points about the disease were:

1.   Taking leaves with very small diseased patches, less than 2 cm in diameter, enclosing them in plastic bags in order to impede contact with outside agents, and observing how the disease eventually engulfed the whole leaf.

2.   Observing how, as a newly forming leaf opens up in a plant where the older leaves are infected, the side nearest to the infected leaves becomes infected in turn and slowly develops fungal patches.

Together these experiments allowed the farmers to perceive the interaction between host and pathogen, the fact that several days are needed for Black Sigatoka to become fully visible, and that from the moment it reaches the leaves, the fungus starts contaminating it. Once they understood this, the farmers were able to work out for themselves the importance of carrying out leaf surgery to remove infected patches.

Impact of the course

Once the members of the school understood each set of principles, they immediately started to adopt the related practices and were able to improve both the yield and the quality of the fruit as well as the overall state of their plantations rapidly.

The farmers were also able to articulate the reasoning behind each of the procedures they were following and to explain why they had changed their practices and what had thereby improved. They increasingly found themselves doing so to other farmers who, in passing, observed what appeared to them to be strange practices and stopped to inquire what they were doing. Once their plantations showed visible improvement, more and more passing farmers wanted to learn how this improvement had been brought about and to find out how they could learn these too.

As a result, at the end of the pilot course there was an immediate demand for new groups to be formed and, in fact, there were offers by a number of the members of the original group to act as facilitators for the new ones, not only from men but also from women. This is especially remarkable because male domination is very high in these regions but also because women are hardly considered farmers at all. Nevertheless, one woman has now become a good facilitator and group leader. Several farmers have formed their own groups, organizing them by means of strict disciplinary rules. They are now working together to see if they can realize the same experiments as the pilot course and ask the same kinds of questions of their farmer peers as the original group of agricultural technicians asked of them.

Conclusion

The method used in this school might be termed transformative education – that is to say, the researchers wanted to support farmers to transform the way in which they conceptualised the agro-ecosystem in which they lived and worked and the position both of plants and human beings within it. At the same time, it was hoped that farmers would also transform their way of acting in the world to consider the benefits of analyzing cause and effect for other issues they are confronted with, not only for farming-related questions. Since this method was being applied to a particular question – that of plantain farming – farmers' attention also had to be drawn to concrete issues and, at least at the start, to the solutions scientists had found for dealing with them. Later on, farmers would be able to build on what they had learned in this way to produce their own concepts from which, in turn, the scientists would be able to learn, thus creating a virtuous circle or spiral of learning.

The approach adopted was to first find a way to move the farmers out of their comfort zones. This was done by using photographs and other materials to produce a psychological shock that opened the farmers to questioning their current assumptions. Then they were allowed to articulate their current understanding and reflect on how they came up with this. Next they were provided with an opportunity for group discussion of the new ideas and finally the participants had the opportunity to put this into practice.[11]

This methodology has been successful in bringing farmers to not only fundamentally reconceptualize the workings of their plantations and approaches to producing healthy plants, but also to reformulating their approaches to their agro-ecosystems, as a whole. They have learned new tools for critical reasoning and are able to apply these to different situations. For instance, the farmers can now challenge agrochemical salespeople by analysing their explanations of why and how chemicals will help them improve their crops and showing them up for false, since they now realize that they do much better without insecticides, herbicides, or chemical fertilisers.

The resources needed to do this are significantly less than for the classic FFS, and potentially this method should be able to spread further and to have much greater impact. However, to get it started, it is necessary to have scientists willing to collaborate on developing holistic farming approaches together with their local farmers, as Suárez and her team have done. Furthermore they need to work together with a specialist in transformative education who can help develop the kinds of materials that can produce the steps outlined in the first paragraph of this conclusion, which can lead to major shifts in thought processes and the acquisition of new sets of tools for critical reflection.

The authors have not started working on mass producing the course. However, the current experimentation by the farmers can provide a good start to doing this and they hope that it will be possible in the future to reach the majority of the 5, 000 plantain farmers in the El Carmen region of Ecuador, and perhaps some of those in Uganda too.

Meanwhile, some of the modules developed for plantain (in particular the safe use of pesticides and the principles of IPM) have been applied to courses

---

[11]   See McGonigal (2005) for a discussion of how these steps can result in transformative learning.

for cacao farming where appropriate, by changing the sets of photographs, main practices and diseases to adjust for the different crop. This is proving as successful with cacao farmers as the course described in this chapter was for plantain farmers.

This experience has led to the realization that, once scientists or extension workers have figured out how to develop a course for one crop, they can relatively simply do so for any other. Moreover, this approach can be used for other aspects of farming and natural resource management as it can for non-farm related subjects.

References

Harris, C., Vera, D., Cabanilla, M., Cedeño, J., Barrera, V., Suquillo, J., Crizon, M., Suárez, C., Norton, G. and Alwang, J.(2003): 'Intrahousehold resource dynamics & adoption of pest management practices'

http://www.oired.vt.edu/ipmcrsp/communications/annrepts/annrep03/Ecuador/14ecuadorcomplete.pdf#page=61 (25th December 2006)

IPM (1996): 'The IPM Farmers' Field School: an Indonesian contribution to sustainable agriculture, The Indonesian IPM Program.

McGonigal, K. (2005): Teaching for Transformation: From Learning Theory to Teaching Strategies', *Newsletter on Teaching*, . 14-2, Spring. Center for Teaching and Learning, Stanford University.

Nathaniels, N. (2005): Cowpea, Farmer Field Schools and Farmer-to-Farmer Extension: a Benin Case Study: Agricultural Research & Extension Network, *Network Paper No. 148.*

Orellana, J., Unda, J. and Analuisa, P. (2002): Estudio de Comercialización del Plátano en la Zona Norte del Trópico Húmedo Ecuatoriano: Ecuador: INIAP, UTE, CAC SD, PROMSA. Publicación Miscelánea 113.

Ramirez, R. (n.d.) Understanding Farmers' Communication Networks: Combining PRA With Agricultural Knowledge Systems Analysis, Gatekeeper 66, IIED.

Schmidt, P. (1997): Farmer Field Schools: A Participatory Large Scale Extension Approach, *BeraterInnen News,* 1/97.

Scoones, I., Chibudu C., Chikura S., Jeranyama P., Machaka D., Machanja W., Mavedzenge B., Mombeshora B., Mudhara M., Mudziwo C., Murimbarimba F., Zirereza B. (1996): *Hazards and Opportunities: Farming Livelihoods in Dryland Africa, Lessons from Zimbabwe*. London and New Jersey: Zed Books.

Sissoko, H.T., Sidibé, I., Harris, C. and Moore, K. M. (2003): Evaluation of Farmer Field Schools for Dissemination of IPM Practices, *IPM CRSP Annual Report 2003*, Blacksburg, VA: Virginia Tech:326-329.

http://www.oired.vt.edu/ipmcrsp/communications/annrepts/annrep03/Mali/12Mali%20Y10%20complete.pdf#page=33 (25 December 2006).

Suárez-Capello, C. Danilo Vera, K. Solís; I. Carranza, J. Cedeño, C. Belezaca; Roger Williams, Mike Ellis; Jeff Alwang, George Norton, Colette Harris; Wills Flowers. (2003): 'Development of IPM Programs for Plantain Systems in Ecuador' *IPM CRSP Annual Report 2002-2003*. Blacksburg, VA: Virginia Tech, 374-380 http://www.oired.vt.edu/ipmcrsp/communications/annrepts/annrep03/Ecuador/14ecuadorcomplete.pdf#page=33 (25th December 2006).

Suárez-Capello, C., Danilo Vera; R. Williams, Mike Ellis; George Norton; C. Triviño; Wills Flowers; K. Solís (2001): 'Development of IPM Programs for Plantain Systems in Ecuador' *IPM CRSP Annual Report 2000-2001*, Blacksburg, VA: Virginia Tech: 381-390.

Werner, J. (1993): *Participatory Development of Agricultural Innovations: Procedures and Methods of On-Farm Research*, Eschborn: GTZ/SDC.

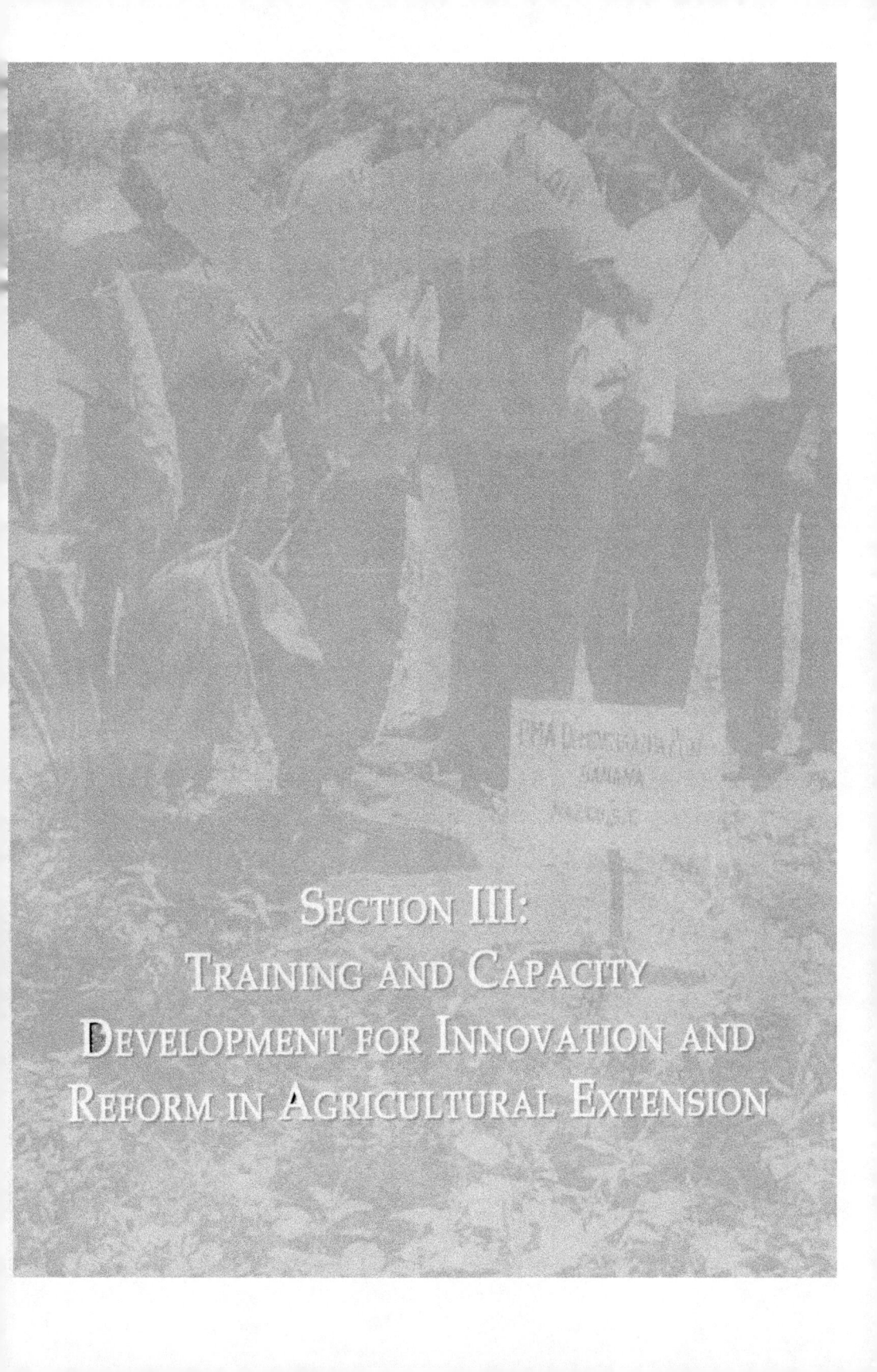

SECTION III:
TRAINING AND CAPACITY DEVELOPMENT FOR INNOVATION AND REFORM IN AGRICULTURAL EXTENSION

CHAPTER 6

The Shift from a Public to a Private Contract Agricultural-Extension System:
Educational Implications of Policy Reforms in Uganda

Jeff Mutịmba, Margaret Najjingo Mangheni
and Frank Matsiko Biryabaho

Introduction

In 2001 Uganda embarked on a process of transforming its public extension system in conformity with the rest of its economic transformations. The public-sector based extension system was gradually phased out and replaced by a contract privatized system implemented by the National Agricultural Advisory Services (NAADS) within a broader policy framework of a multi-sectoral Plan for Modernisation of Agriculture (PMA) involving decentralization, liberalization and privatization. In operational terms, this means that private advisory- service providers work with farmers' organization on contract basis. The major features of the NAADS program approach include private delivery of extension services with public funding; it is demand-driven and farmer-owned; it is a decentralized service; and involves poverty and gender targeting (see Chapter Two for a detailed description of the NAADS program approach).

The shift from public to private contractual extension-service delivery under the NAADS program presents a number of challenges which require human resources with appropriate skills. The system calls for new relationships, knowledge and skills among the stakeholders from the private sector, farmers, farmers' organizations and the public extension system, as they strive to realize the goal of poverty eradication. These new needs create new demands on the agricultural education system, not only in terms of appropriate curricula but also regarding the mode of instruction and training delivery approaches.

This chapter presents results of a series of studies conducted over a period of fours years, between 2001 and 2004, on perceptions of stakeholders, including private contracted agricultural advisory service providers, about new challenges and agricultural education needs brought about by the new privatized contract agricultural advisory system. Based on the analysis, implications are drawn for the design and implementation of educational programs aimed at meeting the diverse needs arising from this system.

Data was collected through a variety of methods including surveys and group and individual interviews. Firstly, data was collected through two consultative stakeholder workshops organized by the Department of Agricultural Extension/Education of Makerere University (DAEE) . Workshop 1, held on 26 to 27 March 2001, before field-level implementation of the NAADS program, was attended by 24 participants, mostly district local government agricultural staff (20) and representatives of a few NGOs (4). Workshop 2, held on 25 February 2002, involved a wider audience with participants from Makerere University, Department of Agricultural Extension/Education (8), NAADS (3), Ministry of Agriculture, Animal Industry and Fisheries headquarters (3), district local government staff (2), NGOs (2), farmer organizations (2) and the agro-industry (1). Secondly, data was also collected from field surveys conducted in three phases. The first phase, between 14 and 22 January 2003, involved NAADS staff at the national secretariat, district and sub-county; NGOs that had served as service providers and farmers' fora in the three trailblazing districts of Kibaale, Tororo and Mukono. Group interviews were used to gather data from members of six farmers' fora involving 49 members in total, whilst individual interviews were used for eight staff of NGOs, from the NAADS national secretariat and nine field staff. In all, six group and 21 individual interviews were conducted. The second phase of the survey was conducted in Kabaale and Soroti districts during the periods 19-20 June and 8-9 August 2003 respectively. Finally, during the third phase, the output of the first two phases was used to design a questionnaire that sought to establish the relative importance of selected knowledge and skills needed by agricultural extension service providers. The third phase covered all six of the NAADS trailblazing districts, Mukono, Kabaale, Tororo, Soroti, Kibaale and Arua and targeted both public (177) and private (68) extension service providers. Thirdly, secondary data from NAADS needs-analysis reports were analyzed.

These were reports of three training needs assessment workshops held in June 2002, whose participants were essentially district local government agricultural staff. Fourthly, data was collected in November 2003 from a survey of 23 graduates of the in-service Bachelor of Agricultural Extension Education (BAEE) degree program of Makerere University[12]. The survey was part of an internal evaluation to identify opportunities for improving the program. Lastly, results of the above assessments were presented to a dissemination workshop of research results on Uganda's National Agricultural Advisory Services (NAADS) held on 12-13 August 2004, which provided further insights and suggestions on the way forward.

Functions of private advisory service providers under NAADS

Development of farmers' institutions, which serve as a platform for farmers to articulate their demands and needs regarding procurement, monitoring and evaluation of services, constitutes one of the key components of the NAADS program process. Accordingly, service providers can be classified into two broad categories, namely those who are engaged in farmer institution capacity building and those providing technology advisory services. The advisory service providers were perceived by the stakeholders as performing nine key functions (Table 6.1) the most important of which was provision of advice to farmers.

---

[12]   The BAEE is a three-year full-time in-service degree program which was launched in 1998 in response to requests from MAAIF for upgrading diploma holding extension staff.

Table 6.1:   *Functions of advisory service providers*

Functions	Data source						
	DAEE ws 1	DAEE ws 2	Farmers fora	NAADS ws	NAADS Staff	NGOs	BAEE graduates
Advising and teaching farming methods	√	√	√	√	√	√	√
Managing technology development sites			√	√	√		√
Providing marketing information			√	√	√	√	√
Building capacity for Farmer organizations	√	√	√	√	√	√	√
Assisting farmers in developing their plans	√	√	√	√	√	√	√
Staff capacity building (retooling)					√		
Documenting activities/ providing reports	√			√	√		√
Monitoring/evaluating programs	√		√	√	√	√	√
Conducting surveys	√				√	√	√
Managing a company							√

Key: ws = workshop.

Whilst the need for advice on farming methods was not new, farmers, in particular, wanted the service providers to put more emphasis on practical demonstrations in their teaching. The service providers are accordingly expected to manage technology development sites set up for farmers to learn from. With the market orientation of the new system, service providers are expected to provide information on marketing. Whilst all the sources consulted saw capacity building among farmers as one of the important functions of service providers, farmers were already beginning to complain about other elements of the mix that were absent. They emphasized that provision of agricultural inputs should go hand in hand with provision of advice about technology and credit and market information. The experience with service providers so far reveals the importance of monitoring and evaluation. Farmers, in particular, felt that service providers should go beyond advice and actually move out to assess the extent to which their advice was being applied.

Challenges under the private contract extension system

The private contract extension system is characterized by a range of new challenges cutting across all levels of program implementation (Table 6.2). The contracted service providers are faced with increased demands for efficiency, time-bound, results-oriented contracts, having to organize their own work and report to farmers. The program managers, on the other hand, face the challenge of developing farmers' capacity to demand and control services effectively, as well as ways of accessing a reliable pool of service providers who can respond to farmers' needs in a timely and appropriate manner.

Table 6.2:   *New challenges under the private contract extension system of NAADS*

New challenges	Data source		
	NAADS staff	NGOs	BAEE Graduates
Increased demand for efficient service delivery	√	√	√
Contracts that are too short to achieve the expected results		√	√
Developing the farmers' organizational capacity	√		
Being accountable to farmers and linking payment to performance	√		√
Political interference		√	
A multidisciplinary approach to solving farmers' problems	√		√
Developing farmers' ability to effectively demand services		√	√
Self-management skills	√		√
Having a reliable pool of service providers who can respond to farmers' needs in a timely and appropriate manner	√		

Coping with increased demand for efficiency and time-bound contracts

The first challenge is meeting the standards of a greater demand for efficient service delivery. Staff of both NAADS and NGOs, as well as graduates of the BAEE program who had participated in the NAADS process felt that there was increased pressure on service providers to deliver quality services efficiently, since they were on contract. This placed new demands on service providers who were used to a less demanding public sector set-up. This challenge was aggravated by the short-term nature of the contracts awarded to service providers, which were perceived to be too short to achieve outcomes such as empowerment of farmers and capacity building, which are of a long-term nature. The feeling was that these processes are slow and difficult, yet the program expects quick results, on the basis of which new contracts will be won or lost.

Coping with farmer-controlled, demand-driven service delivery

Previously, government paid the public extension worker to visit farmers; the approach to extension-service provision was supply-driven, with farmers having little or no input. The new approach, however, is based on a demand-driven principle. Farmers are empowered to decide about the enterprises they want advisory services provided for. Organized in groups and working in a participatory way, the farmers develop work-plans, cost them, draw up the terms of reference for the service providers, look for potential service providers (advertise) and select and contract the service providers. In sum, farmers identify their own needs and specify the terms of service provision for the service providers.

Accomplishing the foregoing tasks requires that the farmers have a certain degree of organizational capacity. NAADS staff felt that, given their lack of experience regarding developing farmers' organizational capacity in aspects such as group formation, group dynamics, and sustaining groups, this aspect of the program approach posed challenges. Another central tenet of demand-driven service delivery is the need to make the providers accountable to farmers. Farmers are expected to participate in awarding contracts and monitoring/supervision of contracts, hence service providers must be accountable to farmers. The challenge is changing a public-sector oriented attitude, which involves service providers being accountable to government supervisors, and forging a link between payment and performance.

Self-management as opposed to being supervised

Under the public extension setup, government managed and supervised extension staff. However, under the privatized system, staff require a new set of skills relating to self-management so that they can perform their functions independently. The new skills are not so much related to technical knowledge, but have more to do with operational efficiency, commercial orientation and targeting various client groups.

Availability of a reliable pool of service providers

The success of the contract extension system depends on the availability of a reliable pool of service providers from which farmers can draw. The development and availability of this pool depends, in turn, on supportive government policies and responsive agricultural education curricula at the relevant institutions.

Competencies needed by the private agricultural advisory service providers

While the studies identified some competency gaps in areas that have traditionally been part of the extension workers' job, such as training skills, there are a number of competencies that are new and certainly unique to a privatized extension system. The latter pertain to those competencies that enable service providers perform independently and cope with the unique challenges of a private setup. Key among these are business management skills and managing personal effectiveness. Under the new system, extension staff are required to form their own business firms, bid for contracts and operate more or less without outside supervision.

Table 6.3 lists a wide range of knowledge and skills identified by stakeholders as being essential for private service providers under NAADS if they are to perform their jobs properly. All the sources consulted emphasized the need for thorough knowledge of the specific technologies involved, together with issues such as environmental concerns, poverty alleviation, and gender and equity goals. In other words, they emphasized the need for a holistic and systems approach to technology promotion and use. They expected the technology advisors to form multi-disciplinary companies since it would be difficult, if not impossible, for individuals to have the range of knowledge and skills associated with all the fields of expertise.

Chapter Six – Public to a Private Contract Agricultural-Extension System

Table 6.3   Competence needs of NAADS service providers as identified by stakeholders

Knowledge/Skill Areas Identified	Data Source							
	DAEE w/s 1	DAAE w/s 2	Farmers	NAADS w/s	NAADS staff	NGOs	Graduates	Disse w/s
Holistic technology-specific knowledge and skills	✓	✓	✓	✓	✓	✓	✓	✓
Rural development		✓	✓					
Natural resource management				✓	✓	✓	✓	✓
Environmental issues		✓			✓			
Livelihood issues					✓			✓
Poverty alleviation					✓	✓	✓	✓
Technology development/testing			✓	✓	✓	✓	✓	✓
Community mobilization/group development	✓	✓	✓	✓	✓	✓	✓	✓
Local languages		✓	✓		✓	✓		✓
Knowledge of local community and customs		✓	✓		✓	✓	✓	✓
Sensitivity to and analysis, of gender/minority groups, HIV/AIDS issues	✓	✓	✓	✓	✓	✓	✓	✓
Participatory methodologies (learning, planning, implementation, monitoring, evaluation, training)	✓	✓	✓	✓	✓	✓	✓	✓

Table 6.3 (contd.)   *Competence needs of NAADS service providers as identified by stakeholders*

Knowledge/Skill Areas Identified	Data Source							
	DAEE w/s 1	DAAE w/s 2	Farmers	NAADS w/s	NAADS staff	NGOs	Graduates	Disse w/s
Communication/facilitation skills	✓	✓	✓	✓	✓	✓	✓	✓
Training program development, training	✓	✓	✓	✓	✓	✓	✓	✓
Adult learning and teaching	✓	✓			✓	✓	✓	✓
Extension methods	✓						✓	
Monitoring and evaluation	✓		✓	✓	✓	✓	✓	
Company formation and registration					✓	✓	✓	
Lobbying and negotiation	✓	✓	✓	✓	✓	✓	✓	✓
Writing skills (contract proposals/reports)	✓			✓	✓			✓
Computer application skills	✓			✓	✓		✓	✓
Contracting					✓		✓	
Conducting/managing demonstrations			✓		✓	✓	✓	
Entrepreneurship/innovation		✓			✓		✓	
Agribusiness including cost/benefit analysis		✓		✓	✓	✓	✓	✓
Business/financial management	✓	✓		✓	✓	✓	✓	✓

Table 6.3 (contd.)   *Competence needs of NAADS service providers as identified by stakeholders*

Knowledge/Skill Areas Identified	Data Source							
	DAEE w/s 1	DAAE w/s 2	Farmers	NAADS w/s	NAADS staff	NGOs	Graduates	Disse w/s
Business/professional ethics					✓	✓		
General management (human and other resources, time, planning, contract management)	✓	✓		✓	✓	✓	✓	✓
Marketing and market information				✓	✓	✓	✓	✓
Knowledge/information management		✓						✓
Coordination/networking skills	✓	✓			✓	✓	✓	
Adoption and diffusion processes	✓							
Research	✓				✓	✓	✓	✓
Leadership	✓	✓			✓	✓	✓	✓
Conflict resolution and crisis management		✓			✓		✓	✓
Policy and legal issues				✓	✓	✓		
Positive personal attributes			✓					
Procurement								✓
Reflective learning frameworks								✓

Difficulties experienced in forming companies were anticipated, given that agricultural advisors had no experience in operating as companies. There was therefore a need for training in company formation and operation, including related policy and legal obligations. The companies have to operate on commercial and business principles. The service providers therefore need skills in all forms of business, financial and human resource management. Unlike in the public service, where everything was provided, service providers now need to know how to prepare financial proposals and budgets, including costing, for their operations. They need to know how to account for funds and adhere to audit requirements, they need to know how to obtain and manage information, and therefore they need computer skills.

Unlike in the public service, where employment was guaranteed, service providers have to be on the lookout for opportunities all the time. This calls for lobbying and negotiation skills, as well as knowledge on contracting and managing contracts, including monitoring and evaluation.

Regarding qualifications of service providers, farmers prefer diploma and first-degree holders to highly qualified scientists. "We don't want doctors (Ph.D). They come and go and we don't see them when we want them", one farmer said. They wanted service providers who were able to operate at grassroots level; who were able to work and live among the communities.

Apart from the relevant knowledge and skills, farmers in particular emphasized that the advisory service providers should have experience. "They should not come to learn from us", said one farmer. This is likely to pose problems for young professionals, unless they team up with experienced practitioners. Farmers also emphasized the need for proficiency in the local language. "They should not ask for clarification all the time", remarked another farmer. In a country with at least 38 languages/dialects (Gakwandi, 1999), this is likely to confine service providers to those areas where they can speak the language. This might mean farmers will not have access to expertise not available in their locality.

Apart from knowledge and skills, farmers said they looked at personal attributes like discipline, keeping time, humility, integrity, self-drive, good personal appearance and friendliness, cooperative and respectful relations with farmers, and commitment to work, tolerance, controlled temper, transparency and political neutrality. Farmers had had negative experiences with advisors who were not conscious of time, arrived late for meetings and expected farmers to wait. Politics is a sensitive topic in Uganda and because of farmers' different political backgrounds, they did not want advisory service providers to "mix agriculture with politics". Corrupt tendencies in awarding contracts were beginning to emerge and were seen as a potential threat to effective service delivery. Farmers therefore wanted the service providers to be transparent.

Table 6.4 below indicates the knowledge and skills required by service providers to perform the different tasks as identified by farmers and other stakeholders consulted.

Table 6.4 *Knowledge and skills required for each key task*

Key Tasks	Required Knowledge/skills
Advising and teaching farming methods	Technical knowledge/skills on specific technology involved rural development, natural resource management, environmental management, gender analysis, training, adult learning/teaching methods, extension methods, planning/monitoring/evaluation, agribusiness
Managing technology development sites	Research and demonstration skills.
Providing marketing information	Marketing, entrepreneurship
Building capacity of farmers' organizations	Community mobilization and group dynamics, local languages, local community and customs, gender/minority group sensitivity and analysis, adult learning and teaching methods
Assisting farmers in developing their plans	Participatory methodologies, communication and facilitation skills, designing interactive learning processes for different clientele – sensitivity to and analysis of gender/minority groups
Staff capacity building	Company formation and business management, computer skills, project proposal writing, human resource management, negotiation, lobbying, advocacy, engaging in multi-disciplinary teams
Documenting activities/providing reports	Writing skills, computer skills, information and data management, financial accounting
Monitoring/evaluating programs	Participatory monitoring and evaluation methodologies
Conducting surveys	Research skills

Implications for agricultural education

As the range of anticipated knowledge and skills generated by various sources indicate, the three issues most likely to have a bearing on agricultural education are the private nature of service delivery, the short contract term and planning horizon, and the fact that the human power pool from which the agricultural advisory system will mostly draw its service providers and trainers is composed of personnel previously trained for a work setting different from the one they are now required to operate in. A discussion of the implications of these issues follows.

Privatization

The privatization of agricultural service delivery has two major implications for agricultural education. First, participation in training will be linked to the benefits of such training. In this setting, individuals and organizations will only be involved in training opportunities and/or activities for which there are potential benefits to the particular individuals/institutions. The sort of free-riding common to publicly organized and publicly-funded training will be less common. Second, is the related fact that the opportunity cost of service provider training will be considerably enhanced. Either the organizations for which the individuals work will compensate the latter, in spite of time lost by the organizations while employees are away on training, or the individuals will forego compensation they would have received had they remained at work. Both of the above have implications for the period of time can be spent away from the work-station and hence the duration and quality of the training programs.

Regarding duration, unlike the public service extension system which could invest in long-term training, private advisory services providers are unlikely to invest in long-term training: a) because they will lack resources for such programs; and b) because of the short term contract (hence planning) horizon they are likely to operate under. They will, therefore, be unable to benefit from the existing agricultural education programs that require long residential periods. Accordingly, if the agricultural education institutions are to play a meaningful role in the human resource development of the private advisory sector, they need to develop custom-made programs for this new and emerging cadre of professionals. One option is designing modular education programs that students can follow without leaving their places of work for long periods. Such programs would of necessity be practical and job-oriented, with identified sets of job accomplishments constituting modules that count towards the award of professional qualifications. In order to enhance the practical aspects, the training institution could adopt approaches such as internships and utilization of farmers and other experienced practitioners as skills trainers to supplement efforts of academic staff, who may have insufficient practical experience. Students would be brought together for short periods, say several weeks, of intensive theoretical orientation and inputs that would be examinable. Students would also complete some theoretical modules through distance learning. In addition, students who successfully complete other specialized training offered by the industry in which they are engaged, would receive credits towards their professional certificates or are exempted from related modules. Under this arrangement, private service providers would be encouraged to update their knowledge and skills periodically, such that, over time, they accumulate sufficient credits for a qualification.

Regarding quality of programs, privatization is likely to exert pressure on educational institutions to offer demand-responsive programs. This will require their transformation into more flexible and responsive learning organizations conducive for quick-action learning. Such an arrangement would benefit from effective mechanisms for linking with stakeholders, notably employers and graduates, for exchange of information and obtaining feedback about programs and emerging training needs.

The short contract term and planning horizon

Reference has already been made to the short planning period dictated by the nature of service-provider contracts. Just as important, however, is the anticipated enterprise turnover necessitated by the design of NAADS. In a bid to avoid spreading resources too thinly at the grassroots level, NAADS currently requires that each sub-county level farmers' forum selects and focuses its planning activities on a maximum of six enterprises for each planning period. One major result of this new arrangement is that traditional linear production systems such as training models are no longer relevant. This implies that curricula will have to be reviewed and redesigned to ensure that pre-service training for the average college or university graduate is geared toward equipping him/her with a critical mass of essential knowledge and skills. In selecting the content areas to cover in this pre-service training, particular attention will have to be paid to knowledge and skills that are prerequisites for graduates to operate reasonably autonomously and/or enhance their own capacity, whether through personal initiatives or structured training. More integrated training as opposed to specialization would be needed at the undergraduate level. The profession should also aim to provide technical backstopping to service providers. This can, for example, be done through short-term, non-certificate, tailor-made courses or field books/fact sheets/manuals/brochures on specific topics.

Retooling current public extension employees

As may be deduced from the range of knowledge and skills cited by the various stakeholders consulted, the advent of NAADS has meant that over and above the roles and tasks traditionally performed by extension agents, the service providers will be expected to perform several new functions. The current pool of agricultural extension professionals, therefore, will require upgrading training programs to enable them function effectively under the new system. Particular attention will, for example, have to be paid to such process-related areas as socioeconomic and gender analysis, participatory approaches, group dynamics, communication, as well as monitoring, evaluation and quality assurance of services. The transfer of learning acquired in such areas to the world of work usually requires attitude change among the individuals concerned. This is a process that can be made easier if these individuals are operating in a supportive environment provided by managers and peers. There is, thus, a need to maintain the momentum generated by training by creating fora for comparing notes on the direction, pace, strengths and constraints of any initiatives resulting from such momentum.

Retooling employees of agricultural educational institutions

Regarding staff of agricultural educational institutions, there will be a need for retraining in the knowledge and skill areas identified by a training needs assessment, but with a special focus on training methodology. Not only will staff need the necessary skills to enable them to integrate subject matter in accordance with the current holistic inclination in the agricultural and rural development world, but they will also have to re-orient their thinking so as to go about their training activities with the right mind-set. The existing internal capacity notwithstanding, linkages will have to be developed and maintained with relevant departments/institutions both within and outside formal agricultural educational institutions, if timely mobilization of the requisite human resources is to be accomplished.

Conclusion

The shift from a public-sector based to a private contract extension approach in Uganda has resulted in new functions, work environments and challenges for both private and public extension service providers. These changes call for a new set of competencies, which introduce new demands on the education system. It is evident that an agricultural education system that services a private contract extension system effectively should have some unique characteristics, the key ones being flexibility, dynamism, responsiveness to the environment and an integrated business-oriented holistic agricultural and rural development curriculum. If extension reforms are not accompanied by corresponding reforms in educational institutions, capacity-related challenges which will undermine the potential impact of the former are likely to arise. Concurrent interventions aimed at reforming the curricula and policies of the supporting educational institutions, in the private and public sector, are necessary for realization of the full potential of extension reforms.

References

Gakwandi, A. (1999): *Uganda pocket facts. A companion guide to the country, its history, culture, economy and politics.* Kampala, Uganda: Fountain Publishers.

Kibwika, K., Mutimba, J.K., Karuhanga-Beraho, M., Semana, A.R. and Matsiko-Biryabaho, F. (2002) *Proceedings of a stakeholder consultative workshop.* 25th February 2002.

MAAIF, (2000). *National Agricultural Advisory Services program. Master document of the NAADS taskforce and joint donor group.* Entebbe: Author.

Mangheni, M.N., Mutimba, J. Matsiko, F.B. (2003): Responding to the shift from public to private contractual agricultural extension service delivery: educational implications of policy reforms in Uganda. *Proceedings of the 19 annual conference of the Association for International Agricultural and Extension Education,* 8-12 April 2003.

MAAIF (Ministry of Agriculture Animal Industry and Fisheries) (1998). *Project implementation completion report of the Agricultural Extension Project CR 2424-UG.* Entebbe, Uganda: December 1998.

NAADS (National Agricultural Advisory Services). (2002): *Report on the training needs assessment for Mukono and Kabale districts,* 12 June 2002.

NAADS (National Agricultural Advisory Services) (2002): *Report on the training needs assessment for private sector institutional development in the NAADS program for Soroti, Tororo and Arua districts*, 19 June 2002.

NAADS (National Agricultural Advisory Services) (2002): *Report on training needs assessment for private sector institutional development in the NAADS program for Kibaale district,* 26 June 2002.

Mangheni, M.N., Okiror, J. and Mutimba J. (2004): *Proceedings of the dissemination workshop of research results on Uganda's National Agricultural Advisory Services (NAADS).* 12-13 August 2004. Department of Agricultural Extension Education, Faculty of Agriculture, Makerere University.

Sseguya, H. and Mutimba J. (2004): *Internal Evaluation Report for the Bachelor of Agricultural Extension Education Degree Program*, Department of Agricultural Extension Education, Makerere University.

Rivera; W; W.M. and Zijp, W. (eds). (2002): *Contracting for Agricultural Extension: International case studies and emerging practices.* New York: CABI Publishing.

CHAPTER 7

Challenges of responding to demand in agricultural education: Experiences of a demand-led agricultural extension training program at Makerere University, Uganda

Jeff Mutimba and Paul Kibwika

Introduction

In the mid-nineties, the government of Uganda directed all ministries, including MAAIF, to reduce staff as part of the structural adjustment program to reduce public expenditure. In MAAIF, staff with diplomas and below faced the threat of retrenchment, which would adversely affect service delivery as most of the extension agents at the grassroots level had no university degrees. In response to this threat, MAAIF requested Makerere University to develop a degree program to upgrade its staff with diplomas to degree level. With the support of the Sasakawa Africa Fund for Extension Education (SAFE), Makerere University developed a three-year full-time BAEE program for mid-career extension professionals, and launched it in October 1998.

For its part, SAFE had experience in supporting universities in the region to develop responsive, farmer-focused, formal continuing education programs for mid-career agricultural extension professionals (Zinnah and Naibakelao, 1999; Zinnah, Steele and Mutocks, 1998)[13]. From its experience, SAFE developed a model for programs whose characteristic processes were

---

[13] The SAFE initiative has expanded from one modest pilot program in Ghana in 1993 to 12 fully established program at colleges and universities spread across East and West Africa.

systematic needs assessment and curriculum revitalization; practical-oriented curriculum and *partnership.* Below is a description of these processes and how they were applied in the case of the BAEE program at Makerere. Thereafter, the challenges of implementing the program are discussed.

Putting the processes into practice
Needs assessment and curriculum revitalization process

To take into account the interests of stakeholders and to initiate partnerships, the SAFE curriculum revitalization process involves six essential steps. These steps provide a flexible conceptual framework for the development of curricula, evaluation and reforms. The framework is not a blueprint, but it is intended to act as a guide to planning curriculum reform in agricultural extension (Mutimba, Zinnah and Naibakelao, 2003).

Step 1: *Informal dialogue among key stakeholders*

This first step usually involves informal dialogue among key actors at selected universities and development organizations – especially the ministries of agriculture. This step aims to induce a joint critical reflection on existing agricultural extension programs, with a view to identifying the key training needs of mid-career professionals. In Uganda, the dialogue mainly involved MAAIF, Makerere University and SAFE in 1995.

Step 2: *Clarifying a common vision*

In view of the outcomes of the first step, the second step involves clarifying the vision for a more responsive extension training program. The vision is based on its relevance as perceived by the stakeholders. The stakeholders then begin to give broad indications of curriculum focus and strategies for its implementation. For the Ugandan case, the program was justified by the eminent retrenchment of the majority of extension staff, which would adversely affect extension service delivery to the small-holder farmers. The BAEE program was then seen as a strategy to save the diploma holders from retrenchment.

Step 3: *Agricultural extension training-needs assessment*

To gain an in-depth understanding of possible content for a responsive curriculum, a formal training-needs assessment is normally carried out by the host university in collaboration with SAFE. The assessment is intended to identify job-based competence gaps – a process that generates both quantitative and qualitative data from a wide range of stakeholders, including employers and extension staff. Formal surveys and informal discussions are the main methods applied to generate information. The capacity (in terms of staffing, facilities and infrastructure and resource materials) of the host university to implement the program is also assessed. Consequently strategies for dealing with the emergent challenges are devised. For Uganda, the formal needs assessment was not conducted; firstly because of the urgency of the program and secondly, the need was upgrading field staff to a degree level to save them from retrenchment. The focus was therefore on getting the program up and running as soon as possible.

Step 4: *Workshop for building consensus with stakeholders*

A workshop involving key representatives of the stakeholders discusses the findings of the needs assessment and works out strategies for sustaining the program. The workshop provides a platform for further dialogue among stakeholders to revisit and reach consensus about the program vision; generate more stakeholder input into curriculum contents; agree on program requirements and criteria/conditions for selection of candidates; and develop mechanisms for long-term partnerships. The partnerships are vital for resource mobilization (both human and financial) and continuous reorientation of the program to meet the changing needs of the stakeholders. For the BAEE program, the stakeholder workshop was held in 1996, and consensus was reached about the nature of the program and a shared vision was further elaborated.

Step 5: *Development of a responsive curriculum*

In discussions about curriculum revitalization, the issue of academic rigour always surfaces. Academic rigour simply means depth of disciplinary subject matter content as perceived by the academic staff and not necessarily the requisite competencies of graduates as perceived by stakeholders. University administrators and staff are usually cautious about launching new programs that are non-traditional or out of the mainstream disciplinary context. Regarding duration, compared to the traditional academic full-time programs, full-time mid-career programs are usually shorter. This is achieved by streamlining the curricula so that no time is spent on non-essential courses and activities and by utilizing all vacations as learning periods for fieldwork and practicals. The program takes into account the fact that candidates present with foundation training in agriculture from their diploma programs and years of field experience. The BAEE program was therefore three years – one year less than the conventional B.Sc. Agriculture programs.

Step 6: *Establishing a strong network among institutions and agencies*

Maintaining and strengthening networks is intended to create an enduring and shared commitment among partners. These partnerships provide mutual benefits through sharing resources, experiences, talents and opinions. One mechanism for maintaining partnerships is through feedback workshops with stakeholders, including students' employers. In workshops, the students present their learning experiences to acquaint stakeholders about the progress and focus of the training programs. Such workshops provide opportunities for university lecturers, students and other stakeholders to exchange experiences and views on the program. In addition, representatives from partner institutions and agencies are encouraged to participate in tours for the purpose of updating themselves on the progress of the program and sharing lessons learnt. This is particularly useful given the risky and innovative nature of such non-traditional ventures. However, due to challenges that are discussed later in this chapter, the partnerships in the BAEE program remained generally weak.

Practical-oriented curriculum

The full-time in-service degree model emphasizes practicals, hands-on laboratory work, problem-focused courses and field-based enterprises. Experiential learning (learning by doing) is at the foundation of the program as it seeks to buttress the practical experience of agricultural extension professionals to enable them to deal with the challenges of agricultural development. As part of their training, students together with their employers, farmers and researchers, develop action-learning projects known as supervised enterprise projects (SEPs) which they implement at their work places for periods ranging from 6 to 8 months. The SEPs aim to solve real-life problems in the field of extension. The students implement the projects under direct supervision of their employers while academic supervisors visit the projects during each vacation to provide on-the-spot guidance. The SEPs provide an opportunity for co-learning between students, their employers and university lecturers in real-life situations. SEPs provide unique and rare opportunities for academic staff to assess the relevance and effectiveness of their teaching and to identify other opportunities for learning. The projects, also known as supervised extension projects, provide a mechanism for actualizing and strengthening partnerships between the university and employers through their joint efforts to assist in solving problems in the community.

Partnership

As the program is developed with employers as partners, the SAFE model requires that this partnership be nurtured and maintained throughout implementation, with the roles of the partners clarified. Normally a memorandum of understanding (MoU) to this effect is signed between the employer and the university.

Challenges posed by the in-service degree program for mid-career public service extension professionals at Makerere

The irony of qualifications versus competencies

The BAEE program emerged as a response to problems of structural adjustment that targeted staff retrenchment as a means of reducing government expenditure. The staff with lower academic qualifications were more vulnerable, as academic qualifications emerged to be the criterion for retrenchment. The interest of the demand side (MAAIF) was therefore to enable their extension staff with diplomas to acquire degrees and not necessarily competence. The urgency of such a demand forestalled the need for the systematic training-needs assessment elaborated above. The result was that the program started with a curriculum which was less than ideal. Revising the curriculum took several years, given the bureaucratic procedures of the university in revising curricula.

The challenge is how to respond to demands for education. Demand can be expressed in different ways; in this case, it was expressed in terms of qualifications which, in itself, questions the purpose of education. Behind qualifications is the assumption of competence. If universities are to develop programs that will stand the test of time, they need to process the demand beyond what is commonly expressed as qualification, to unravel the underlying competencies that are assumed by qualifications. Doing so requires engagement with stakeholders and conducting processes of a strategic nature in an environment that does not exert undue pressure on any of the partners. Responding to the desire for higher qualifications, as in this case, only addresses the symptoms of a problem rather than the cause. Universities then begin to act like "fire fighter brigades" instead of putting in place measures to prevent the outbreak of fire. A quick fix to problems will not go far; universities need to adhere to processes of developing competency based programs that, in the longterm, save time and other resources.

Aiming at a moving target: the challenge of responding to dynamic policies

Soon after the BAEE program was launched at Makerere, the government's new decentralization policy shifted the responsibility for agricultural staff recruitment, development and deployment from MAAIF to the local governments (districts). The policy shift automatically changed the partnership arrangement. The university was then faced with the challenge to start new negotiations with districts to provide scholarships for their staff. One partner (MAAIF) multiplied into more than 50 new employers (local governments) to negotiate with the University. Obtaining commitments from the local governments for scholarships was difficult, as training was not among their top priorities. Districts also lacked resources to invest in staff development, and they were, in any case, faced by other more pressing problems. The university then invested in vigorous marketing of the program to the districts. The demand had now reversed – the university was now demanding instead of the local governments. The university found itself facing a dilemma of how to sustain the program with small student numbers, as few of the potential candidates could access scholarships.

While pursuing negotiations with the local governments, another policy to privatize the agricultural extension services through the National Agricultural Advisory Services (GOU, 2001) was implemented. This policy anticipated that the extension staff would be laid-off (de-layered) from the public service and would be expected to form private companies to provide advisory services on a contract basis with farmers. Given the low interest of the private sector in investing in long-term training, the challenge for the university to secure scholarships for candidates for the program became even bigger. Scholarships for some of the candidates already enrolled on the BAEE program were cancelled by the local governments as they could no longer justify spending on a service that had been privatized. Such students had to pay their way through the program. Enrolment on the program fluctuated from year to year, at times the numbers did not justify the existence of the program.

Some university staff suggested lowering the admission requirements in order to raise the critical numbers to sustain the program, which would mean that the focus would shift from quality to numbers. The design then would not fit and probably a redesign would be necessary. The shift in policies also changed the employers for the potential candidates of the mid-career program. Adapting the program was more like aiming at a moving target, given the many changes that were taking place. This posed a great challenge for the inflexible university system. Such a dynamic multi-stakeholder development environment calls for transformation of higher education institutions into more flexible, dynamic, client-oriented learning organizations that can cope with change proactively.

Doing new things in the old way

The BAEE program was intended to be innovative based on action-learning processes. Since the candidates of the program bring with them considerable work experience, it would be valuable to utilize their experience as a learning resource on the basis of which new perspectives could be introduced, to deepen learning. Action learning would engage both the students and lecturers, to learn together to explore new horizons of knowledge and new frontiers of innovation. This signifies a shift from the teaching paradigm to the learning paradigm. As Ison (1990) puts it, universities reinforce the teaching paradigm by describing their purpose and function as "custodians" and "preservers" of knowledge – which creates the image of knowledge as a "commodity" that can be "stored" or "warehoused" and then dispensed or given (by lecturers) to recipients (students). But Dreyfus and Wals (2000) emphasize that we may never understand the problem until we start to actually implement some potential solutions and that, without the ability and willingness to act, it is impossible to participate in or, rather, to contribute to a democratic society. For universities to understand the problems they are expected to assist in solving, they have to take part in implementing possible solutions to problems in society (Kibwika, 2006).

In this view, the SEPs were designed to be action-learning projects where the students would engage with the lecturers, other stakeholders and community to solve real-life problems. This was intended to happen right from the start, throughout the program, by utilizing the holidays for SEPs activities. However, while lecturers were expected to play a key role in facilitating reflective processes that generate the learning, they were not familiar with action/learning/action research paradigms themselves. The dean of the Faculty of Agriculture at Makerere once remarked "... I have staff who have never milked a cow but they are teaching animal science." The tendency, therefore, was to fall back to the "comfort zone" of lecturing. In terms of teaching approaches, there was no difference between the conventional programs and the innovative BAEE program. Consequently, the tendency was to combine BAEE students with the conventional B.Sc. program students for most of the courses, although this was not meant to be so.

The key issue to consider here is the competence of university lecturers to facilitate innovative learning processes. Exposing academic staff's ignorance is part of the process of engaging in and facilitating action learning (see Revans, 1997) which threatens the hierarchical "knowledge-power" relations between lecturers and students. Action learning is threatening to the lecturers because it "disempowers" them of their "expert" power. Thorpe, Taylor and Elliot, (1997) point out that, from an individual staff perspective, the problems associated with learner-centred learning include the loss or devaluing of skills, fear of not being the expert, need for new skills and changes in student expectations. Dealing with such insecurity is a critical challenge for the shift from teaching to learning. To promote innovation in universities, the entry point would have to be to reorient the academic staff to innovative ways of teaching, doing research and consultancy as well as changing mindsets. To be able to change others, academic staff need to change themselves first. It would be naive to imagine that academic staff can produce a new quality of graduates without changing their own ways of teaching. The competence required here is not necessarily about getting higher degrees like Ph.Ds, though it is desirable. It is also about being able to do things differently. Competence of staff to facilitate learning is often taken for granted, especially in universities, and yet it is the core foundation for innovation.

Reorienting mindsets of students

The students come to university with a mind-set of going back to school. Their previous experiences of school and the environment in which they find themselves have a big influence on what they consider learning to be. Interactive or action learning was as new to the mid-career students as it was to the other students, and even lecturers. Garrant (1997) refers to action learning as a process for reforming of organizations and liberation of human vision within organizations. In this case it is liberating the students from the teaching or "lecture" that they were accustomed to, to new learning approaches. This liberation does not happen instantly, it is a process. Deviation from the lecture method is often seen as departure from the "normal". The learning approach that requires total immersion of the whole person also requires dealing with epistemological aspects. There has to be a deliberate effort to re-orient the students to new ways of learning, which includes consistency in application of interactive learning approaches.

The BAEE students were no exception, also expecting the lecturer to deliver knowledge while they were on the receiving end. Their role then would be to reproduce the knowledge during examinations. This is the perception that shapes the entire education system, and is even stronger at universities than at lower levels of education because lecturers tend to limit their examinations to reproducing what they have taught. The incentive to explore learning beyond what is delivered in lectures is absent. After all, the same lecturer teaches, sets and grades the examinations.

Innovative learning programs have to have built-in mechanisms for targeting change in students' mindsets towards learning. The value of interactive learning can only be appreciated and sustained if the entire process of learning, including assessment, encourages interpretation and logical analysis of issues in a systems context. While the BAEE students seemed to enjoy interactive learning approaches that explore and link their experience to real-life professional challenges, they would often prefer to have things the easy way; that is, "give us what you have and we give it back to you in the exam". Engaging students in processes that enable them to discover knowledge and its application is more challenging to the mind. So if I can get a degree the easy way, why bother? Again, this can be related to the external system that values qualifications more than competence. In other words, to encourage innovative learning in universities, there is a need for a new culture that inspires students to develop competence rather than just acquire qualifications. This implies questioning the current purpose of education and building new values and aspirations through education.

Responsiveness of the university system

While universities are expected to champion change, they are the slowest to respond to change. The university system and setup does not allow flexibility for quick response to the changing environment. Curriculum reviews take long to pass through the bureaucratic processes, to the extent that changes in curricula are usually implemented when some of them have already lost their relevance. For example, although the BAEE curriculum was reviewed and revised, the process of implementing the revised curriculum stalled. University regulations stated that a curriculum had to pass through at least an entire cycle before proposals for its review can be considered – a full cycle means at least three years. Again, from the time a curriculum review is initiated, experience shows that it takes another 2-3 years to have a revised curriculum approved by the University Senate.

In addition, processes of curriculum review remain largely expert driven, with lecturers determining the curriculum content based on their background training. Feedback from other stakeholders, such as employers and students, are often dismissed as having no theorital foundations. For the BAEE Program, consultative meetings with the students and employers, for example, proposed that some courses be deleted from the curriculum because they were irrelevant and that others be included, but such proposals met with resistance from the lecturers. The challenge is to establish institutional structures and mechanisms that enable change to flourish when it is still relevant. This calls for a re-thinking of university structures (including rules and procedures) and culture, to allow for quick responses to stakeholder needs. In addition, the value attached to input of other stakeholders into curricula needs to go beyond rhetoric.

Resources and the hype of sustainability

Universities today operate in a competitive environment similar to corporate organizations. Ironically, they are expected to be innovative and do more with less resources (see UNESCO, 2003). This is against common knowledge, namely that for innovations to succeed through evolving processes of refinement, resources are required. Innovations usually start small as experiments – in this case, the program starts with a few students. At this stage, it is impossible to make such programs self-sustaining in terms of funding. At the same time, university funding is reducing and all new programs are required to run on a cost-recovery basis for sustainability. In this regard, the BAEE was established as a private program and therefore expected to generate funds for its operations. While SAFE provided some resources to kick-start the BAEE program, such external funding is not sustainable and neither does it cover all the facets of the program. The University needs to support strategic innovations that will influence change, to make it more relevant to development needs.

Due to resource (including human) constraints, BAEE students were merged with the conventional programs for most courses thereby diminishing its uniqueness. Other activities that accompany innovation, like student field supervision, also require heavy investments in vehicles, fuel and staff allowances, which the program could not meet on its own. Such constraints stifle the good intentions of innovation at universities, that might otherwise have provided models for change.

Conclusion

Changing development and social contexts leave universities with no option but to change their training, research and outreach services to respond appropriately to emerging challenges. While universities are criticized for being slow to respond to new development challenges, the task is a daunting one. The case of the BAEE described in this chapter demonstrates clearly that universities can actually respond to new educational demands, albeit with limitations. Public universities are constrained by ever-reducing funding from governments, yet they are expected to present innovations which require enormous resource investment. The support from SAFE illustrates that, with financial and technical support, universities can respond appropriately and also change their way of conducting business. While the commercial approach being adopted by universities like Makerere may stimulate innovations, it also has the potential to damage the quality of academic programs. However, the bureaucracy of universities and their rigidity regarding change from the conventional way of doing business will have to give way to adaptive management at all levels of the university.

The inappropriateness of the university response is also compounded by the level of articulation of demand and the circumstances under which that demand is expressed. Undue pressure is often placed on universities to respond to short-lived problems which compromise the processes for developing more sustainable competency-based programs. This is largely due to overdependence on potential resources from the demand side. For universities to present quality innovations, they require resources under their control that are not conditioned by the demand. In addition, building strong partnerships with stakeholders, especially employers and/or industry, is fundamental to relevance and sustainability of innovations at a university. A critical factor for the universities is to build the competences of their staff to facilitate and manage innovative educational programs. Equally critical is a system that values, reinforces and rewards innovation, otherwise all the efforts exerted on behalf of innovations can quickly fall back into the conventional thinking "basket". *"If we always do what we have always done, we will always get what we have always got!"* A transformative view of education is overdue at universities.

References

Dreyfus, A. and Wals, A.E.J. (2000): Anchor points for integrating sustainability in higher agricultural education. In W. Van de Bor, P. Holen, A.E.J. Wals and L.W. Filho, *Integrating concepts of sustainability into education for agriculture and rural development,* Environmental Education, Communication and Sustainability, Vol. 6. Peter Lang. pp.73-91.

Garrant, B. (1997): The power of action learning. In M. Pedler, (ed.) *Action learning in practice,* Third Edition, London, Gower Publishing, p.p. 15-29.

GOU(Government of Uganda) (2001): National Agricultural Advisory Act. Acts Supplement No. 9 to the Uganda Gazette, No. 33 Volume XCIV, dated 1 June 2001.

Ison, R.L. (1990): Teaching threatens sustainable agriculture, *Gatekeeper Series No. 21,* International Institute for environment development, Sustainable agriculture program, London.

Kibwika, P. (2006): Learning to make change: developing innovation competence for recreating the African university of the 21st century. Wageningen, Netherlands: Wageningen Academic Publishers.

Kline, S.J. and Rosenberg, N. (1998): An overview of innovation, in R. Landau, N. Rosenberg (eds)., *The positive sum Strategy: harnessing technology for economic growth,* Washington: National Academic Press, p.p. 275-305.

MAAIF (2000). *National Agricultural Advisory Services.* Master Document of the NAADS Task Force and Joint Donor Groups. Entebbe, Uganda.

Mutimba, J., Zinnah, M.M. and Naibakelao, D. (2003): Toward reorienting agricultural education in Africa to ensure quality and relevance: ten-year experience of Sasakawa Africa Fund for Extension Education. In A.B. Temu, S. Chakeredza, K. Mogotsi, D. Munthali and R. Mulinge (eds). 2004. *Rebuilding Africa's capacity for agricultural development: the role of tertiary education.* Reviewed papers presented at ANAE Symposium on Tertiary Agricultural Education, April 2003. ICRAF. Nairobi, Kenya: 43-51.

Revans, R. (1997): Action learning: its origins and nature. In M. Pedler (ed.), *Action learning in practice,* Third Edition, London: Gower Publishing p.p. 3-13.

Rip, A. (1995): Introduction of New Technology: Making Use of Recent Insights from Sociology and Economics of Technology, *Technology analysis and strategic management* 7: 417-431.

Thorpe, R., Taylor, M. and Elliott, M. (1997): Action learning: in an academic context. In: M. Pedler (ed.), *Action learning in practice,* Third Edition, London: Gower Publishing. p.p. 145-172.

UNESCO (2003) Synthesis report on trends and developments in higher education since the World Conference on Higher Education (1998-2003) Paris: UNESCO.

Zinnah, M.M. and Naibakelao, D. (1999): Bringing African universities into development: The SAFE program at the University of Cape Coast. In S.A. Breth (ed.), *Partnership for rural development in sub-Saharan Africa,* 65-75. Geneva: Centre for Applied Studies in International Negotiations.

Zinnah, M.M., Steele, R.E. and Muttocks, D.M. (1998): From margin to mainstream. In *Training for agricultural and rural development,* 16-28. Economic and Social Development Series No. 55. Rome: FAO

SECTION IV:
SELECTED EMERGING ISSUES IN AGRICULTURAL EXTENSION SERVICE DELIVERY

CHAPTER 8

Agriculture and the Environment:
Implications for Extension in Uganda

Monica Karuhanga Beraho and Emmanuel K. Beraho

Introduction

While agriculture continues to be a major source of livelihood for a greater majority of people in Uganda and has positively contributed to the food security of the country, the sector remains the major cause of environmental degradation (National Environment Management Authority [NEMA], 2004). It is important to note that the resultant environmental degradation arising from the inappropriate use of agricultural practices also in turn limits agricultural production and productivity. Consequently, agriculture becomes a cause and a victim of environmental degradation. This scenario is likely to lead to a progression of a vicious cycle of continued use of unsustainable agricultural activities, increased environmental degradation, as well as increased vulnerability of agricultural-based livelihoods that are likely to have limited capacity to implement recommended agricultural practices. It is therefore impossible to talk about modernization of agriculture and eradication of poverty without talking about sustainable use of natural assets and ecosystems that underpin the agriculture. In recognition of the relationship between natural resources and people's livelihood, the World Bank (2002) succinctly stated: "for people to thrive, assets must thrive". United Nations Development Program (UNDP) report (2005) further strengthened this by observing that delivery on the Millennium Development Goal (MDG) 7, devoted to ensure environmental sustainability, will result in delivery on all other MDGs, including MDG 1 aiming at eradicating extreme poverty and hunger. The Human Development Report (UNDP, 2003) underscores the need to focus on the needs of the people whose livelihoods depend on natural resources and environmental services because of the strong relationship between poverty and the environment.

It has been noted that the poor are particularly affected by the degradation of natural resources and loss of biodiversity, not only because they depend on them for their sustenance and income, but because of the fragility and marginality of their lands (United Nations Capital Development Fund, 2004). World-wide 1.3 billion people live on marginal land (that is, land prone to soil erosion, nutrient loss, semi-arid areas, mountainous, wetlands) (World Bank, 2003). Inhabitants of fragile ecosystem make up a large proportion of those classified as living in extreme poverty. In view of the above, the importance of natural and environmental resources in national development cannot be over emphasized. Therefore ensuring that they are well managed is vital. Because agriculture is the predominant activity in developing countries like Uganda and the sector that utilizes the biggest portion of land and natural resources, the centrality of the agricultural sector in environment management cannot be emphasized enough.

This chapter seeks to highlight the effect of agriculture on environment; existing opportunities and challenges for integrating environmental messages in extension; and the role of extension in promoting sustainable use of Natural Resources for agricultural development.

Effect of Agriculture on the Environment

Environmental degradation is a result of the dynamic inter-play of socio-economic, institutional and technological activities. Environmental changes may be driven by many factors including economic growth, population growth, urbanization, rising energy use and transportation, as well as the extensification and/or intensification of agriculture. The impacts of agriculture on the environment can be attributed to, among others, inappropriate farming techniques, over exploitation of land and water resources, intensive use of chemicals, encroachment on fragile and protected ecosystems as well as inadequate and/or inappropriate extension services. Even where extension services exist, messages on sustainable environment management have been very limited. Furthermore, a greater majority of agricultural extension service providers have not effectively integrated recent developments and national strategies in environmental management of relevance to agriculture. Consequently, natural resources and ecosystems have deteriorated and their capacity to support agriculture have reduced tremendously. The decline in agricultural productivity has resulted in poverty among farming communities, despite advances in agricultural research. Therefore sustainability should be a requirement that must be met by the agricultural sector. Sustainability must not be traded off in favour of short-term agricultural benefits. Major effects of agriculture on the environment are discussed below.

Effect of Agriculture on Soil

One of the most important consequences of agriculture on the environment is soil erosion, which is the principal manifestation of land degradation in Uganda. NEMA (2002) reveals that 46.5% of the land area of 19 districts is affected by soil erosion. Slade and Weitz (1991) estimate that soil erosion accounted for the loss of 4-10% of the gross national income and represented about 85% of the total annual cost of environmental degradation. Yaron and Moyini (2003) estimate the economic cost of soil erosion to be 11% of GDP. Soil nutrient loss studies have estimated soil nutrient loss to amount to about $626 US million per annum. This translates into a debt of $52, 000 US million per year and a per capita soil erosion debt of $210 US to be paid by future generations. Soil erosion is caused by surface run-off or wind where vegetation cover has been removed through clearing of land for agriculture, poor cultivation practices, deforestation, bush burning and overgrazing. Kabale, Kisoro and Mbale are the worst affected districts. The uncontrolled run-off of excessive rainwater from the mountains and hilly slopes in these districts causes gullies and landslides (in the most severe form), with equally serious consequences for soil productivity. In many areas of the country, particularly those that are densely populated and where continuous cultivation is practiced, rill and sheet erosion is evident. Soil erosion, in addition to degrading the land, also degrades water bodies. Siltation poses a major threat to Uganda's water bodies and wetlands (see Plate 8.1). A high sediment load of up to 1 ton/day has been measured in rivers located in mountainous regions (NEMA, 2002).

Other features of soil degradation are salinity, soil compaction and pollution. Soil salinity, which makes soil unsuitable for crop growing, has been reported in the Semiliki Valley south of Lake Albert. Areas under perpetual irrigation are also said to be prone to soil salinity. Soil compaction has been associated with areas where mechanized farms exist (such as Jinja, Mukono and Mpigi districts and the Kigumba area in Masindi district) and where ox-ploughing is practiced (because of shallow ploughing depth) such as in the districts of Kumi, Soroti, Katakwi and Lira. Localized soil pollution due to agricultural activities has been reported in areas where there is intensive use of agrochemicals, for example, the flower farms near Lake Victoria and the livestock farms using acaricides (NEMA, 1998).

If Uganda's soil nutrient loss is taken into account, it becomes apparent that Uganda's development is not on a sustainable path and corrective interventions in the agricultural sector are required urgently. It takes about 500 years to form 25 mm of soil under agricultural conditions (Pimentel, Huang, Cordova and Pimental, 1996).

Effect of Agriculture on Biomass

Forests, trees and woodlands are important resources and play multiple ecological, economic, social and cultural roles. Uganda's forest cover, has however decreased drastically. FAO estimated the forest cover to have been about 10.8 million ha in 1890, or 52% of Uganda's surface area (MUIENR, 2000). This has shrunk to 5 million ha or 24% of the land surface area. In 2000, the rate of deforestation[14] in Uganda was estimated to be 55, 000 ha per year[15] (Forestry Department, 2000). Major factors contributing to deforestation are encroachment, land conversion to agriculture (see Plate 8.2), unsustainable harvesting, urbanization, industrialization and institutional failures (policies and laws) (NEMA, 2001).

The opening up of new land to cope with increasing demand for food as a result of population increases has contributed to deforestation of woodlands and a few forest patches on public and private land (NEMA, 2001). This has resulted in loss of vegetation cover and reduced crop yields. Expansion of agriculture on previously forested steep terrains has had serious environmental impacts in the districts of Mbale, Kapchorwa, Kisoro

---

[14]   Kampala, Jinja, Luwero, Iganga, Mubende, Mbale and Tororo districts have the highest percentage of degraded forest area, while ditricts with the least degraded areas include Nebbi, Soroti, Rukungiri, Moyo, Masindi, Kabale, Kalangala, Bundibujo and Bushenyi (NEMA, 1998).

[15]   Other studies estimate the rate of land clearance to range from 70,000 to 20,000 ha per year (MFPED, 1994). It is estimated that land use change or deforestation would be equivalent to about 200,000 ha annually (Forestry Department, 2002).

and Kabale, with soil erosion leading to landslides, siltation of rivers and lakes and loss of water catchment basins. The promotion of non-traditional agricultural export crops (NTAEs) is also said to have led to increased opening up of new land for agriculture (NEMA, 2001).

Overgrazing and poor agricultural techniques such as slash and burn (see Plate 8.4) have also contributed to unnecessary clearing of woody vegetation. Such practices have been found to be prevalent in districts like Sembabule, Rakai, Mbarara, Nakasongola, Kiboga and Luwero (District Environment Reports). Additionally, encroachment on natural forest reserves (though to a limited extent) by cattle keepers (in districts of Kiboga, Mubende, Nakasonola, Luwero, Soroti and Iganga) and by cultivators (in Kiboga, Bundibujo and Iganga districts) has also been reported (NEMA, 1998).

One of the key factors that has caused forest degradation in the past has been the inadequacy of forests laws and legislation, particularly in the context of decentralized governance and environmental management. It is hoped that recent revisions of the relevant policies (if well implemented) will make them more effective for managing forest resources in the country. The Plan for Modernization of Agriculture (PMA) which, among others, aims to improve the efficiency of the rural farmer MFPED and MAAIF, 2000), if well implemented, also offers opportunities for slowing conversion of forested areas to agricultural uses, as productivity on existing farms is improved.

Effect of Agriculture on Wetlands

Wetlands have long been known to provide a buffering capacity against pollution, flooding and siltation. They are known to provide such functions as recharge of underground water, modulation of the climate, and serve as a source of seasonal pastures. Agricultural activities have contributed significantly to the degradation of wetlands (see Plate 8.3, 8.4 and 8.5). NEMA (2005) estimated that a total of 2, 376 km2 of wetland have been reclaimed in Uganda, mainly for agricultural purposes. Wetlands, particularly those with shallow water, have been put under intensive cultivation of crops such as sugarcane, sweet potato, yams and eucalyptus. Quite often, these wetlands have been badly managed. For example, the reclamation by Kakira Sugar Works of the Mutai swamp forest, where River Kiko originates, for purposes of growing sugarcane, has led to the loss of wetland's capacity to clean the waste water (contaminated) originating from the Kakira Sugar processing plant (NEMA, 2001). The districts of Jinja, Iganga and Pallisa have the highest percentage of modified wetlands in Uganda (NEMA, 2005). Rice production has resulted in conversion of 60% of wetlands in Iganga district. Kabale district has converted 74% and Jinja about 80% of its wet lands. In addition to compromising the environmental functions of wetlands, agriculture also poses potential public health problems. Most people who work in wetlands do so unprotected, exposing them to health problems such as bilharzia. This should concern the agricultural extension officer as much as it does the health worker.

Effects of Agriculture on Drylands/Rangelands

Poor agricultural practices have contributed immensely to the degradation of drylands. Globally, it is estimated that 60% of drylands are degraded, resulting in an estimated annual economic loss of $42 billion worldwide. It is also estimated that desertification is affecting an additional 6 million hectares each year. This global picture reflects the reality in Uganda. Continued dryland degradation is therefore, clearly, a threat to national economy and the welfare of the people. In Uganda, more than 29 districts, with a combined population of about 10 million people, live in dryland areas (plate 8.6 and 8.7). It would therefore be illogical to believe that the national poverty eradication drive can succeed without a clear and practical plan to prevent degradation of drylands and to promote their productivity through improved and highly-targeted agricultural extension.

The importance of drylands notwithstanding, they seem to have received the least attention in this country by extension agencies. The key driving force behind degradation of drylands is a nexus of poverty, rapid population growth, agricultural activities, especially overgrazing, uncontrolled bush burning by pastoralists and crop farmers (plate 8.8), wanton destruction of vegetation cover, deforestation and other poor agricultural land-use practices (plate 8.9 and 8.10). Consequently, destruction of drylands in Uganda has contributed to a decline in agricultural production and food security, a decline in levels of major water bodies, heavy loss of livestock, poverty, poor livelihood and social conflict.

Effect of Agriculture on water resources

The whole of Uganda's water resources is part of the Nile Basin shared by 10 countries. Over the last two decades, the quality of surface water in Uganda has been deteriorating. (NEMA, 2001), which has, in turn, affected the water quality in the Victoria Nile. Agricultural run-off and industrial and municipal waste-water discharge into the lake from riparian countries is considered one of the factors causing changes. The situation is said to be exacerbated by catchment degradation, intense cultivation of land, poor agricultural practices such as cultivation up to the river bank/lake shore and lack of soil erosion measures, and conversion of wetlands around the Lake Victoria to other uses (plates 8.10 and 8.11).

The consequences have been siltation of rivers, lakes and wetlands. Sedimentation is a significant non-point water pollutant in Uganda. It causes water to become cloudy, creating an unsuitable environment for many aquatic species, reduces the water holding capacity of rivers, lakes and streams and sometimes requires expensive dredging. Prevention of siltation due to inappropriate farming remains a challenge to the agricultural sector. Most ecosystems in Uganda are threatened by siltation. Wetlands in many districts such as Kabale and Kisolo are as degraded by siltation due to poor farming practices uphill as they are degraded by agricultural conversion. Lake Nakivale in Isingiro district is massively silted as a result of inappropriate agricultural activities in the surrounding hills, to the extent that some parts have gradually converted from lake to wetland to dryland. River Manafa in Mbale and Hatari in Kapchorwa are similarly affected.

In addition to sedimentation, chemical contamination of both surface and underground water resources is a problem associated with fertilizers and pesticides used in agricultural production. As the population increases and more food is required, the use of fertilizers, pesticides, and other agrochemicals is likely to increase. If their application is not well guided, it could be an important water pollutant. Nitrate contamination of groundwater pose potential health risks, especially to newborns and infants younger than 6 months. Phosphorus is a potential surface water pollutant with significant effects on aquatic resources. Manure contains nitrogen, phosphorus, inorganic salts, organic solids and micro-organisms. All these are potential contaminants of both surface and groundwater. The run off from poorly designed feedlots, manure storage facilities and poor application in the fields can be a direct source of surface water contamination. Some parts of Kampala, Iganga, Wobulenzi and Kasese have been identified as prone to groundwater pollution (NEMA, 2001).

Pesticides affect water quality through run-off or leaching into streams. Pesticides may affect non-target plants, fish, birds and beneficial insects. Mishandling of pesticides in storage areas, during mixing and spills contribute to environmental degradation by contaminating soil and water resources. Health concerns include applicator exposure to dangerous concentrations of pesticides and to the public's long-term exposure to pesticide residues in food and water. The challenge is to maximize the benefits of sound pesticide management while reducing environmental risks. Pesticide applicators who read and follow instruction on pesticide labels can minimize risks to human health and the environment, a challenge to the largely illiterate Ugandan farming community. Promotion of integrated pest management, record keeping, crop rotation, selective planting and harvest dates, pest-resistant crop varieties, biological control and avoiding farming very close to water bodies could be some strategies extension agencies can promote to limit the impact of pesticides. Other chemicals commonly used on farms such as paints, wood preservatives, oils, to mention a few, should also be disposed of properly because they are potential pollutants of water and soil, in addition to posing health hazards to people who use them.

Climate change

In addition to the effect of agricultural activities on the climate due to wetland degradation and biomass degradation, agriculture is also a major producer of greenhouse gases (http://www.panda.org/about_wwf/what_we_do/ policy/agriculture_environment/commodities/beef/envir.../index.cf). Globally, ruminant livestock produce about 80 million metric tons of methane annually, accounting for 22% of global methane emissions from human-related activities. While this is probably not yet a big problem to Uganda, it is important for extension agencies to know these facts because as agriculture is modernized, animal production will be developed further and methane production will increase. Uganda is a party to the convention on climate change. Uganda is therefore obliged to take all necessary measures to mitigate activities that affect climate change, for example, by reducing emissions of greenhouse gases. Sustainable agriculture can contribute significantly to this effort and therefore extension should be brought on board.

Plate 8.1   *River Manafa is characterized by severe siltation, erosion of river banks, and cropping up to the edge of river bank. Its quality is reduced, as water is brown. See extent of gravel in centre foreground (Mbale District).*

Plate 8.2   *Deforestation. Land conversion for agriculture has led to the loss of vegetation cover and exposed soil to erosion (Kibale District).*

ALL PHOTOGRAPHS, COURTESY OF NEMA, UGANDA.

Plate 8.3   *Wetland degradation leads to loss of environmental functions (Jinja District).*

Plate 8.4   *Wetland Conversion for crop production. Poor farming practices of slash and burn (Kabale District).*

ALL PHOTOGRAPHS, COURTESY OF NEMA, UGANDA.

Plate 8.5   *Wetland Conversion for Livestock production (Kabale District).*

Plate 8. 6   *Human settlement in a rangeland (Kumi district).*

ALL PHOTOGRAPHS, COURTESY OF NEMA, UGANDA.

Plate 8.7   *A Munyankore herdsman with his cattle. Rangelands are a very important source of forage for livestock for rural communities (Kumi District).*

Plate 8.8   *Burning exposes the sandy soils in rangelands to erosion (Masindi District).*

ALL PHOTOGRAPHS, COURTESY OF NEMA, UGANDA.

Plate 8.9 *Deforestation exposes soil to erosion and leads to forest degradation (Kibale District).*

Plate 8.10 *Charcoal burning in a rangeland. Diversification of income sources may have negative environmental impacts (Kumi District).*

ALL PHOTOGRAPHS, COURTESY OF NEMA, UGANDA.

Plate 8.11   *An example of poor farming practices. A maize crop has been planted up to the river bank leading to an increased danger of water pollution through fertilizer and pesticide in surface run-off.*

Plate 8.12   *Banana plantation. About 68% of Ugandans depend on agriculture as their main source of income (Kapchorwa District).*

ALL PHOTOGRAPHS, COURTESY OF NEMA, UGANDA.

Promotion of biogas production is another strategic intervention area that extension and relevant departments ought to consider. Not only would this rid the environment of gases like methane, but the gas could also be used for cooking and lighting (thus reducing the pressure for fuel wood on forests and woodland while partly solving the problem of limited rural electrification).

Environmental effects associated with processing of agricultural products

The effect of agriculture on the environment is not limited to the production phase in the field only. Agriculture is broader than farming. It also embraces processing and marketing of agricultural products. To this end agriculture-related industrial pollution has added a new dimension to the threat to lake ecosystems. For example, over 40% of municipal solid waste results from agricultural products (NEMA, 2005). The leading polluting industries, the majority of which are agro-related, include breweries, textiles, sugar factories, soft drink, dairy, oil and soap, food processing and leather tanning industries. Most of these industries are concentrated in Kampala, Jinja, Entebbe, Masaka, and Mbarara. Waste from these industries is discharged directly into the major surrounding water bodies (Lake Victoria, Katonga River and Nile) with potential negative effects on humans and wildlife. For example, effluent from tanneries processing hides and skins are usually high in chromium and biochemical oxygen demand levels.

Environmental effects on social harmony

Decreased capacity of ecosystems to support agriculture due to misuse is one of the reasons why people migrate from one region to another. Usually this results in social conflict and further degradation of fragile ecosystem because migrants usually occupy agriculturally marginal areas and fragile ecosystems such as wetlands, forest reserves, protected areas, riverbanks and lakeshores. Therefore the consequences of environmental degradation due to poor agricultural practices are not limited to degradation of natural resources and the associated reduction in agricultural productivity. Environmental degradation often results in social conflict as well. In Uganda, for example, the occupation of the Teso region wetland by pastoralists from other regions could be attributed partly to degradation of natural resources in areas where these pastoralists originate. Their migration to occupy the Teso wetlands has not only resulted in the degradation of the wetlands, but has also resulted in conflict with the local people. The same applies to massive migration of people to Kibale district from other parts of country.

The role of extension in environmental management

Concern for the environment and how it could be reconciled with the need to increase agricultural production is an important issue for agricultural extension. Extension must not emphasize increased agricultural production exclusively, but must ensure that the increased production is achieved in a sustainable manner. This requires effective and efficient use of environmental assets. It is becoming increasingly clear that environmental issues are very important to the agricultural sector, therefore agricultural workers' proactive role in environmental affairs is a prerequisite for improved agricultural productivity and sustainability. The primary role of extension is to minimize the negative environmental impacts of agriculture while ensuring adequate production of food. This can be achieved if agricultural extension practitioners realize that they have a leading role in the management of environmental matters associated with agricultural practices and that negative consequences to the environment as a result of agriculture should, as far as possible, be planned for and funded by the agricultural sector in collaboration with other relevant stakeholders. The following are some of the roles agricultural extension agencies could play.

1. *Sensitize and train farmers and other stakeholders about recommended environmentally friendly agricultural practices and environmental legislation related to agriculture:* Extension workers could play a frontline role in ensuring that farmers apply sustainable agricultural practices and comply with environmental legislation related to agriculture. While environmentally sound practices may involve cash outlays, extension agencies should emphasize the wisdom in paying now to prevent degradation of assets that support agriculture, rather than paying more later as a result of decreases in agricultural output and the need to restore degraded environmental assets and

ecosystems. The obligation of farmers to ensure intergenerational equity should also be emphasized. Extension agencies should therefore generate and disseminate environmental information related to agriculture to maintain a constant awareness about issues relating to the environment in agriculture and to facilitate decision-making and learning. In order to integrate environment information in extension messages effectively, it is important that extension workers acquaint themselves with and promote existing environmental legislation and guidelines. For example, guidelines and legislation related to access and use of fragile ecosystems such as wetlands, mountainous and hilly areas, riverbanks and lakeshores are now in place.

2. ***Analyze relevant policy and advise policymakers:*** Extension agencies can, and should, proactively influence policy and legislation at all levels in order to enhance sustainable agriculture. Extension professionals should have the capacity to analyze the potential environmental consequences of agricultural policies and legislation and advise on integration of the necessary environmental safeguards and their implementation modalities. Not only should extension analyze proposed policies, they also have a role to originate policy aimed at enhancing sustainable agriculture. Extension workers at sub-national level, for example, could play a crucial role in advising local authorities on the formulation of policies and byelaws aimed at enhancing sustainable agriculture at household and community levels. Such laws would include, for example, bylaws related to access and the use of local resources, for instance community wetlands, with a view to preventing deleterious agricultural activities. Extension would also be vital in organizing communities to protect and use local resources prudently for agricultural purposes. This may entail organizing communities to establish and observe community agreements and norms with regard to access and use of community resources.

3. ***Continuous monitoring of environmental effects of agricultural activities and taking timely action:*** Extension workers at local level interact with farmers more than their counterparts in environmental agencies do. Therefore the role of extension workers in monitoring environmental degradation associated with farming cannot be overemphasized. Extension workers should therefore be involved

in monitoring the degradation of environmental resources due to agricultural activities and work with relevant sectors to address the issue in a timely manner, before it becomes irreversible.

4. ***Play a liaison role:*** Extension agencies should play a key liaison role between all institutions and stakeholders interested in the promotion of sustainable agricultural production and environmental management. For example, they can link farmers with researchers. In this capacity extension agencies are is a comparatively good position to spearhead efforts to develop and test sustainable agricultural technologies and practices. In addition, extension agencies can also link farmers and other stakeholders to environment and forestry department staff for more specialised advice, as well as to NGOs, CBOs and other relevant institutions.

Challenges

Addressing environmental concerns in agriculture poses the following challenges:

1. ***Limited knowledge among farmers regarding the environment:*** The agricultural sector is threatened by chronic environmental degradation mainly as a result of poor agricultural practices and techniques due to low environmental awareness among farmers. This is compounded by the fact that farmers in many rural areas lack or have inadequate access to agricultural extension services in general and those that promote sustainable use of environmental resources in particular.

2. ***Limited funding for agricultural extension:*** The long-standing inadequate funding of the agricultural sector in general and agriculture extension-service delivery in particular has been largely responsible for the limited penetration by extension agents of farming communities to impart knowledge and skills. Therefore the challenge faced by the extension service to promote sustainable agriculture in the context of inadequate funding is a reality they have to contend with.

3. ***Cost of implementation of environmentally sound practices:*** Farmers tend to regard application of environmental safeguards as a type of production costs. The challenge is to convince them to invest minimally in sustainable agricultural practices today to enhance

their agricultural productivity in the future. Continuing to apply environmentally harmful practices will inevitably lead to a decrease in the capacity of environment resources to support agriculture. Farmers will have to therefore spend more, economically and socially, to survive in future. In the light of massive rural poverty, the issue sometimes that certain households, particularly female, grandmother, and child-headed households, lack the resources to implement recommended practices. While land degradation is known to be one of the major effects of agricultural activities, poorer farmers usually find it difficult to implement appropriate soil and water conservation practices because their operates are highly labour intensive. Nonetheless, it is high time that both the poor and rich realize that sustainable utilization of environmental resources has cost implications. Rural farmers must be made aware that the effects of environmental degradation are likely to affect the poor more than other social category.

4. *Limited knowledge about environmental issues among extension workers:* While extension agencies are expected to spearhead the effort to address environmental issues in the agricultural sector, few extension workers are trained to handle this task adequately. Agricultural training institutions have not adequately integrated environmental issues in their curricula to ensure that extension personnel are environmentally 'literate' by the time they complete their studies and are deployed in the field. If extension agencies are to play their environmental role effectively, this needs to be addressed.

5. *Poor attitude of extension agencies:* Ensuring that extension workers embrace their role in addressing environmental aspects associated with agriculture will require attitudinal transformation among extension workers. Currently the dominant thinking is that environmental management is the preserve of environmental agencies such as NEMA and the district environment officers and that integration of environmental issues in agriculture presents extension personnel with additional work. Environmental issues related to agriculture need to be considered as part and parcel of agricultural activities and processes undertaken by the agricultural sector. For example, it does not make sense for a farmer to obtain information on improved seed varieties from an agricultural extension worker and then wait for advice from

the environment officer to tell him/her not to grow this seed along a river bank or in a wetland. Environmental issues can be incorporated in routine extension work without additional staffing and funding or institutional changes/reorganization. Integration of environmental roles into traditional agricultural extension roles still remains a challenge that requires urgent redress by strategies related to curricula and institutional practices.

6. *Lack of harmonization of agricultural and environment policies:* Agricultural development policies and plans have not been adequately subjected to environmental assessment. This type of assessment will determine environmental consequences of agricultural activities with a view to identifying and incorporating appropriate mitigation measures for minimizing its negative effects. Agricultural extension emphasizes and promotes increased agricultural production at the expense of environmental assets that underpin production. For example, government policies, such as the double crop production campaigns of the 1970s and the promotion of the livestock industry through the establishment of cattle ranches, contributed to large-scale clearing of land (NEMA, 2001). Furthermore, even though the PMA talks of increasing productivity while ensuring sustainable use and management of natural resources, there is a need to evaluate, among other things, the extent to which this has been operationalized. Strategic environment assessment of agricultural policies, plans and activities should be adopted by agricultural agencies.

7. *Developing and strengthening linkages:* The perception that agricultural production is the preserve of the agricultural sector has resulted in weak inter-sectoral linkages between the agricultural sector and other sectors that impact on agriculture. Some of the major determinants of effective agricultural production lie outside the mainstream agricultural sector. Therefore promotion of sustainable agriculture cannot not be realized by the agricultural sector alone, though it should spearhead the effort. Establishing and strengthening linkages and coordination with environmental organizations, the forestry sector, educational institutions, NGOs and CBOs, among others, is extremely important though highly challenging.

Conclusion

The direct impacts of agricultural development on the environment arise from farming activities which contribute to soil erosion, soil depletion, loss of biomass, land salination, water and soil pollution, and health and safety hazards of workers, among others. Commercialization of agricultural enterprises has often been accompanied by overexploitation of land and water resources and increased use of agrochemicals. Intensive agriculture and irrigation contribute to land degradation, particularly through soil compaction, salination, alkalization and waterlogging. Leaching from extensive use of pesticides and fertilizers is an important source of contamination for water bodies. Industrial pollution from agro-industries has added a new dimension to the threat to Uganda's aquatic ecosystems. Intensification of agriculture activities due to population pressure has also been an important cause of land degradation.

Like in other parts of Africa, the problem of environmental degradation is of serious concern to Uganda. In a continent where too many people are already malnourished, crop yields could be halved within 40 years if the degradation of cultivated lands is to continue at present rates (Scotney and Dijkhuis, 1989). The problem becomes more significant when one realizes that agriculture and agriculture-related activities constitute the largest sector of the economy and the major source of income for most people, including the poor. Rising levels of poverty and income inequality[16] present big challenges with regard to poor people's capacity to undertake sustainable agriculture. The high population growth rates of 3.4 percent (UBOS, 2003) create an urgent need to enhance agricultural production and productivity to satisfy the increasing demand for food (NEMA, 2001). Agricultural production, being a land-based activity has, without a doubt, potentially adverse effects on the environment; however the effects can be mitigated, and therefore the challenge is effective integration of appropriate mitigations in agricultural production processes. At this time both public and private extension institutions ought to broaden their scope of work

---

[16] Income poverty increased from 34% to 38% between 2000 and 2003, and inequality as measured by the Gini coefficient rose markedly from 0.39 to 0.43 (MFPED, 2004).

beyond transfer of agricultural technology and consider integrating other crosscutting issues, such as the environment, into ongoing agricultural extension programs, with a view to ensuring sustainable rural livelihoods.

As part of a future strategy, the curriculum of agricultural training institutions should integrate environmental issues. However, for those agricultural extension personnel already in the field, in-service training on environmental issues of relevance to agriculture needs to be conducted. Among other qualities, extension personnel should have the capacity to apply environmental assessment techniques to agricultural policies, plans and projects/activities. This is crucial for predicting the impact of agricultural policies on the environment, monitoring the state of the environment in relation to agricultural activities and implementation of sustainable agricultural practices. Developing the capacity of extension personnel to apply environmental assessment techniques to agricultural activities has the potential to ensure sustainable agriculture. There is a need for guidelines/checklists to guide environmental assessment of agricultural activities and to ensure that extension personnel are exposed to their application. Development of environmental indicators to assist in monitoring environmental changes related to agriculture and measuring the performance of extension in addressing environment concerns in agriculture would equally be useful.

References

Forestry Department (2002): National Biomass study Technical Report. Kampala, Uganda: Ministry of Water, Lands and Environment.

MUIENR (Makerere University Institute of Environment and Natural Resources), 2000. National Biodiversity Data Bank Report 2000. Kampala, Uganda: Makerere University.

Ministry of Finance, Planning and Economic Development (1994): Statistical abstracts 1994. Kampala, Uganda.

Ministry of Finance, Planning and Economic Development (2005): Poverty Eradication Action Plan: Kampala, Uganda.

NEMA (National Environment Management Authority) (1998): State of the Environment Report for Uganda, 1998.

NEMA (National Environment Management Authority) (2001): State of the Environment Report for Uganda 2000/2001.

NEMA (National Environment Management Authority) (2002): State Of the Environment Report for Uganda: Kampala, Uganda: Access Reprographics, .

NEMA (National Environment Management Authority) (2005): State Of the Environment Report for Uganda draft 1: Kampala, Uganda: Access Reprographics.

Pimental, D., Huang, X., Cordova, A., Pimental, M. (1996): Impact of Population Growth on Food Supplies and the Environment. Paper presented at the American Association for Advancement of Science (AAAS) Annual Meeting, Baltimore, MD. 9th February 1996

Scotney, D.M. and Dijkhuis, F.H. (1989): *Recent changes in the Fertility Status of South African Soils. Pretoria, South Africa: Soil and Irrigation Research institute.*

Slade, G and Weitz, (1991): Uganda Environmental issues and options. Masters Degree Thesis. Unpublished. North Carolina, USA: Duke University.

UBOS (Uganda Bureau of Statistics, (2003): National Household Survey 2003, Report on the labour force survey.

UNDP (2005): Uganda Human Development Report. Linking Environment to Human Development: A Deliberate Choice Kampala, Uganda: UNDP.

UNDP (2003): Human development Report. Millennium Development Goal: A contract among nations to end human poverty. United Nations Development Program YK.

UNCDF (United Nations Capital Development Fund) (2004): Local Environmental Governance and the Decentralized Management of Natural Resources: New York. UNCDF.

World Bank (2002): Reaching the rural poor. Washington, World Bank.

World Bank (2003): World Development Report: Sustainable Development in a dynamic world. Transforming Institutions, Growth and Quality of life. Washington: World Bank.

Yaron, G. and Moyini, Y. (2003): The Contribution of Environment to Economic Growth and Structural Transformation. Report prepared for ENR Working Group for the PEAP Revision Process. Kampala, Uganda: MFPED.

(http://www.panda.org/about_wwf/what_we_do/policy/agriculture_environment/commodities/beef/envir.../index.cf). (accessed 30 November, 2006).

CHAPTER 9

Impact of HIV/AIDS on Agriculture and Rural Livelihoods in Uganda:
Implications for Agricultural Extension Services

Monica Karuhanga-Beraho

History and current status of HIV/AIDS in Uganda

Uganda has been affected by the HIV/AIDS epidemic for almost a quarter of a century. A history of civil war and political conflict, economic collapse in the 1970s and early 1980s, migration, geographical positioning and extreme poverty have worked in concert at different periods in time to form a lethal mixture that either fueled the spread of the disease or increased the susceptibility of certain social groups to HIV (Haddad and Gillespie, 2001). The epidemic started on the shores of Lake Victoria in Rakai district (located in the south-western part of the country), the initial epicentre of the illness. Thereafter, HIV infection spread quickly, initially in major urban areas and along highways. By 1986, HIV had reached all districts in the country, resulting in what is classified as a generalized epidemic. As in other countries in sub-Saharan Africa, the most common means of transmission of HIV still remains unprotected sex with an infected person (84 percent), although mother-to-child transmission has become an important route, as evidenced by the number of children with AIDS at the end of 2002 (UAC, 2003a). It is estimated that about 2 million people were infected by HIV throughout the first 25 years of the epidemic, of whom about 1 million have died and another 1 million are living with the infection today (UNAIDS, 2006). Results from the 2004 UHSBS indicate that just over 6 percent of Ugandan adults are infected with HIV, with the prevalence among women higher than among men (8 and 5 percent, respectively). Moreover, people living in urban areas have higher prevalence relative to those in rural areas (Ministry of Health and ORC Macro, 2006). Nearly 80 percent of those infected with the disease are between the ages of 15 and 45 years, the most economically productive group[17] and often those that tend for families (UAC, 2003a). This is particularly important in terms of future labour availability, given the labour-intensive nature of agricultural activities. Since the beginning of the epidemic, about 2 million Ugandan children (approximately 25 percent of all Ugandan children) have been orphaned by AIDS (UNAIDS, 2006; UAC, 2003a).

Consequently, the HIV and AIDS epidemic has had far-reaching social and economic consequences that have affected individuals of all walks of life and communities nationwide. The impact of the disease has been mainly the increasing levels of morbidity and mortality among women and men. For Uganda, food insecurity, degraded livelihoods, increased vulnerability and adverse socio-economic impacts have in many instances been identified as causes and consequences of HIV and AIDS (MFPED, 2003). At community level, declining agricultural productivity is eminent and the death of prime age adults has imposed unsustainable strains on the extended family structure. The massive number of orphans means that an increasing number of households are headed by children. The micro- and macro-economic consequences are diverse but centre on the loss of critical human capital, reduced industry and private- sector growth as well as diversion of meagre government resources to the health sector for the control and prevention of HIV and AIDS, with negative effects on economic growth and poverty reduction. It is becoming increasingly evident, therefore, that the impacts of the HIV and AIDS epidemic are undermining development initiatives.

Impacts of HIV/AIDS on the agricultural sector and agriculture-based livelihoods

The role of the agricultural sector in national development and the livelihoods of the majority of Ugandans cannot be overemphasized. In spite of steady improvements in the growth of services and industry sectors and a decline of 12 percentage points[18] in agriculture's contribution to GDP over the past decade, the greater majority of people (69 percent) still depend on agriculture as their main source of income (UBOS, 2003). Apart from providing food security and farm incomes, agriculture also supports

---

[17] It has been estimated that Uganda will have lost 14 percent of its agricultural labour force due to AIDS by 2020 (MFPED, 2004).

[18] Agriculture accounted for 38.7 percent of GDP in 2002/3, as compared to 51.1percent in 1991/92.

the agriculture-based industries. For the majority of people in rural areas most of the income opportunities are agricultural, comprising specifically of crop and livestock combinations (UBOS, 2005). As Carney (1998, cited in Haddad and Gillespie, 2001) explains, a livelihood represents the interaction between assets and transforming processes and structures that generate a means of living, all conditioned by the context that individuals find themselves in. In order to appreciate the potential impacts of HIV and AIDS on rural livelihoods, it is important to understand the context within which people live, particularly the vulnerability context, to which HIV and AIDS has now become an important part. The impacts on rural populations, their livelihoods, their farming systems and on food security have been especially severe.

Impacts on farming systems

HIV and AIDS impacts on the farming system are usually a consequence of the long-term effects of the epidemic on households (composition or dissolution) due to HIV and AIDS-related labour loss. Impacts on farming systems therefore include reduction in cultivated land, a decline in crop yields and the variety of crops grown; and changes in livestock (Du Guerny, 2002; Barnett and Blaikie, 1992); a loss of valuable agricultural skills and experience (Hurst, Termine and Karl, 2005; Du Guerny, 2002) as well as rising labour costs due to labour shortages (Hurst *et al.,* 2005). Furthermore, poor agronomic practices due to labour shortages are usually associated with an increase in prevalence and spread of plant and animal diseases (UNDP, 1995). There is evidence suggesting variation in the sensitivity of farming systems to labour loss as a consequence of HIV and AIDS (Barnett and Blaikie, 1992; Barnett, Tumushabe, Bantebya, Sebuliba, Ngasongwa, Kapinga, Ndelike, Drinkwater, Mitti and Haslwimmer, 1995; Rugalema, 1999; Mongi, 2002). Barnett and Blaikie (1992) suggest that farming systems characterized by a low seasonal variation in the demand for labour, a high degree of crop diversity with labour requirements that differ in seasonal peak times or systems with labour supply exceeding demand, are more resilient (or less vulnerable) to HIV and AIDS impacts. They further suggest that farming systems situated in the semi-arid tropics and that have one major cropping season will be more vulnerable to labour loss as a result of HIV and AIDS. Barnett and Blaikie (1992), in their study on Uganda, combine household- level impact studies based on extensive fieldwork with farming-system classification. In mapping HIV sero-prevalence and adaptations to labour loss and other characteristics, they identified 9 (out of 50) farming systems to be vulnerable to the AIDS pandemic in Uganda. The northern Mubende and western Luwero, the northern Hoima, Kabarole, Kasese, Southern Iganga and Tororo districts were classified as under serious threat. While south-western Mubende, north-western Masaka, and southern Hoima districts were said to be under moderate threat. Currently, most of Uganda's farming systems are less vulnerable to HIV and AIDS- related labour loss (Topouzis, 2000).

While it is important to understand impacts at farming-system level, such information needs to be complemented by that obtained from empirical data on how affected households respond in a situation of HIV and AIDS. As it has been argued, it is difficult to predict the impact of HIV and AIDS on the household using farming-system analysis because analysis at this level does not take into consideration inter- and intra-household variations in resources, wealth and food security (Barnett and Blaikie, 1992) or the diversity of livelihoods that farmers exhibit, or the opportunities they have to diversify into activities that may not be highly labour dependant (Haddad and Gillespie, 2001).

HIV and AIDS Impacts on rural livelihoods

Although the impacts on rural livelihoods can be exemplified in concrete terms by considering the household level, it also includes the wider environmental or institutional context that mediates access to resources and hence livelihood options, and which may also constrain the functioning of any given household. An analysis of the impacts of the epidemic in the context of the sustainable livelihood framework can facilitate a better understanding of the dynamics between HIV and AIDS and rural livelihoods. The sustainability of livelihoods depends on how assets are mobilized, managed and enhanced so as to preserve them. Sustainable livelihoods are those that can avoid and resist stress and shocks (such as HIV and AIDS) and are able to bounce back when affected, while unsustainable (vulnerable) livelihoods cannot cope with these changes without being damaged (Niehof and Price, 2001). HIV and AIDS is known to increase this vulnerability. As Haddad and Gillespie (2001) conclude from their analysis of the impacts of the epidemic on five categories of assets, HIV and AIDS strips individuals, households, networks and communities of different forms of capital (human, financial, social, physical and natural) thus reducing their future capacity to cope with other shocks (see Box 1 below). The current trends in income poverty[19] and inequality in Uganda can only depict a probable increase in numbers of households with reduced capacity to cope in the face of overwhelming crises created by the HIV epidemic. Comparative case-study research has shown that different livelihood groups (for example: male versus female household heads; or men and women in male-headed households) exhibit different responses to the HIV and AIDS-related impacts, due to differences in their asset base and respective livelihood strategies. Consequently, as Bebbington (1999) asserts, households with a robust and stable asset base and resources are likely to generate a more sustainable livelihood compared to households that have neither enough assets/resources nor the capability to create or access them. For the majority of people in rural Uganda, most income opportunities are agricultural, comprising specifically crop and livestock combinations (UBOS, 2003). However, women, widows and female-headed households, male youth, households with large families, and people depending on vulnerable sources of income, such as those relying on fisheries, nomads and small-scale farmers growing only one low-value crop, form the most vulnerable social groups in Uganda (MAAIF, 2000). HIV and AIDS are known to increase this vulnerability. HIV and AIDS-induced changes in household composition and structure are of particular importance for availability of household for agricultural production and other income-generating activities and, consequently, the capacity of households to generate sustainable livelihoods.

---

[19]   Income poverty increased from 34 percent to 38 percent between 2000 and 2003, and inequality as measured by the Gini coefficient rose markedly from 0.39 to 0.43 (MFPED, 2004).

Box 1   *Excerpts from interviews conducted during an HIV and AIDS study in Masaka District, Uganda.*

This disease (AIDS) has crippled development in our area. Move around the village and all you will see are old people and orphans. All our young blood, the people who were starting to develop the area, died, leaving us these young children. We can only look at them. We have nothing to offer them. They are being chased from free government-aided primary schools because we cannot afford to buy pencils, books or pay for their lunch at school. What does the future hold for them? Indeed some of them have started robbing us of the little food we have! (Focus group discussion with women, Nyenje, Lwengo, Masaka District, Conducted April 2005, by M.Karuhanga-Beraho).

Responses to HIV have included withdrawal of children from school to bolster a family's ability to provide care and/or maintain its current livelihood strategies, a switch from cash crops to food crops and a decline in the variety of crops grown. It is important to note that, while some of these responses might make sense at the time, they are likely to lead to destitution as they compromise the household's capacity to maintain and generate a sustainable livelihood.

HIV/AIDS Impacts at rural household level

The impacts of HIV/AIDS at the level of the farming household have been summarized in a number of reviews (Barnett and Blaikie, 1992; Haddad and Gillespie 2001; Loevinsohn and Gillespie, 2003). The most immediate impact of HIV/AIDS at the household level has been associated with effects on human capital and/or other household assets. Therefore household level impacts can be summarized as decline in household labour availability, asset depletion and knowledge loss.

Human capital: At the household level, perhaps one of the most critical impacts associated with HIV and AIDS attacks on human capital is labour loss. Infected individuals die prematurely while those living with the disease are rendered less productive or non-productive once the disease emerges. There is also the issue of foregone labour when labour for agriculture is diverted to caring for the sick. Beyond labour loss is another more important loss, that is, loss of the knowledge and experience of the dead individual. The magnitude of impacts is likely to vary for different individuals and households given that the progression of the disease varies from individual to individual, depending on the type of syndrome exhibited, their immune and nutritional status, and ability to access treatment for opportunistic infections and Anti-retro viral drugs (ARVs), among others. According to UNDP (1995), the negative effects of HIV/AIDS on agricultural labour are mostly the result of death of the adult household breadwinner. In addition to reduced agricultural labour, the quality of life of those left behind declines due to worry and increased workload for the remaining spouse or older children as they try to look after the rest of the family.

It is worth mentioning at this point that, although little is documented about the impacts of HIV and AIDS-related child mortality, there are unforeseen social implications regarding the future labour force.

Financial resources: Adult sickness and deaths due to HIV and AIDS have had severe economic consequences for surviving household members. This is because savings and assets have been diverted to meet medical and other related costs (for example, special foods) associated with caring for the sick, as well as funeral costs (UNDP, 1995 and World Bank, 1997 cited in Wakhweya, Kateregga, Konde-Lule, Mukyala, Sabin, Williams and Heggenhougen, 2002). Because of depressed immunity due to the disease, affected individuals are usually advised to improve on their nutrition by including foods high in proteins (eggs, chicken, beef) and vitamins (vegetables and fruits) in their diet. However, proteins constitute one of the most expensive groups of foods that is ordinarily consumed only on festive days in rural areas. Therefore, the impact on household resources of the inclusion of such foods on a more regular basis for an HIV-infected individual cannot be overemphasized. Cross-country analyses reveal that income loss due to HIV/AIDS ranges from 40 to 60 percent in households directly affected (UNAIDS, 1999 cited in Wakhweya *et al.*, 2002) and per person income is about 15 percent lower in orphan households, and property ownership significantly less (World Bank, 1997). If the remaining parent is also sick, which is usually the case, and/ or dies, such households are likely to be left with no labour or economic potential and surviving members have diminished capacity to generate a livelihood or respond to any future shock.

The poor usually rely on informal credit or group-based microfinance products in times of distress. Unfortunately these types of services tend to be spatially concentrated and hence vulnerable to aggregate shocks (Gillespie *et al.*, 2001). Gillespie, Haddad and Jackson (2001) further argue that, even when the epidemic is in its early stages, affected households are less likely to avoid default and hence become less attractive to group-based liability schemes.

Rural households tend to utilize remittance and off-farm income to purchase agricultural inputs and equipment such as fertilizer, improved seed, spray pumps and ploughs for use in farm production (Reardon, Crawford, and Kelly, 1995; and Marenya, Oluoch-Kosura, Place, and Barrett, 2003). However, as noted by Yamano and Jayne (2004), these sources of income are often jeopardized among AIDS-affected households, particularly those that are asset-poor. Cash constraints for increasing farm production are further compounded during illness and after a death, when medical and funeral expenses rise and care-giving by other members reduces their income earning potential as well (Jayne, Villareal, Pingali and Hemrich, 2005). Further evidence from the IP survey (IP & FAO, 2003), shows that in Uganda, the following could be observed in all households affected by HIV and AIDS:

- a proportional decrease in the amount of money spent on farm equipment and agricultural inputs;
- reduced uptake of recommended agronomic practices, such as row and line spacing, appropriate depths, compost and manure making;
- the storage and use of seed for sowing rather than the purchase of costly high-yielding varieties; and
- infrequent hire of tractors for preparing land.

Physical assets: In Uganda, land is the most important physical asset of the poor. Most land is acquired through inheritance rather than purchase (Kwesiga, 1998), and land titles and tenure tend to be vested in men either by legal condition or by socio-cultural norms (World Bank, 1993). In the IP survey conducted in Uganda (IP & FAO, 2003), there was evidence that affected households were three times more likely to sell land compared to non-affected ones. The same survey revealed that affected female-headed households had an average reduction in landholding of 11 percent (0.3 acre), owing to distress sale and the loss of land to relatives following the death of a spouse. In the event of the death of a spouse, Ugandan law allows a widow to retain 15 percent, conversely, in practice this share may be withheld partly because some women do not know their rights but more so because of existing socio-cultural factors. However, labour shortage predisposes households to the risk of losing ownership rights, especially in areas where these are not clearly defined and where land access is closely linked to use (UNDP, 1995; Haddad and Gillespie, 2001; and Wakhweya *et al.*, 2002).

Because land is usually an important livelihood asset for the rural poor, it is the last asset that is disposed of in the case of severe stress (IP & FAO, 2003). Evidence indicates that households first attempt to sell off small animals, and other assets with less impact on long-term production potential (Jayne *et al., 2005,* Cattle and productive farm equipment are sold in response to severe cash requirements, for example, after the death of a key bread-earner within the family (Yamano and Jayne, 2004). Given the fact that most of these are distress sales, these assets are disposed of at a far lower value than they would normally fetch. Asset depletion compromises the future survival capacity of those left behind. As Jayne *et al.* (2005) argue, the cumulative effect of asset depletion and therefore reduced availability of disposable cash to acquire inputs for agriculture may cause a decline in the capability of small-scale farmers to produce and, consequently, the proportion of a marketable surplus from farming.

HIV and AIDS Impacts on the agricultural estate sector

Likewise, private-sector organizations are affected by the epidemic on various fronts. Not only does HIV and AIDS rob them of staff and institutional knowledge, but the epidemic also has significant financial implications. To date, however, the impact of HIV and AIDS on the commercial sub-sector has been less documented than effects on the semi-subsistence sector. This is the case in spite of the fact that, being largely dependent on migrant labour, the commercial sector is highly vulnerable to HIV and AIDS too, as labourers often live on the estate away from their families. They can therefore become an important element in the spread of AIDS since social networks tend to be weaker in such settings. Consequently, rural communities have been found to bear a higher burden of the cost of HIV and AIDS, as many urban dwellers and migrant labourers return to their villages when they become sick.

A study of the impact of the AIDS epidemic on commercial agricultural production in two districts of Kenya revealed that the commercial agriculture sector of Kenya is particularly susceptible to the epidemic and is facing a severe social and economic crisis due to AIDS impacts (Rugalema, 1999). Morbidity and mortality due to HIV and AIDS significantly raise the industry's direct costs (medical and funeral expenses) as well as indirect costs through the loss of valuable skills and experience. In this way the epidemic adversely affects a company's efficiency and productivity, and coping strategies aimed at reducing the costs often prevail over strategies aimed at HIV/AIDS prevention. Therefore, the impacts of HIV and AIDS-related illnesses in agro-estates results in rising costs (due to high medical and funeral costs that exceed budgetary provisions, as well as recruitment and retraining of new staff) as well as in decreasing profits (due to loss of workers and working hours). Reduced productivity due to reduced morale among workers, and other psychological effects, have also been documented

(Rugalema, 1999). In Uganda, key estate crops like tea (in Kabarole, Bushenyi, Mukono and Mubende), and sugar (in Mukono and Jinja) districts, for example, also rely on migrant labour, implying that Uganda's agricultural estate sector is likely to experience similar or even worse impacts, given that HIV prevalence in Uganda has been higher than that in Kenya. Cyclical migration to coffee growing areas such as Mbale, Bushenyi and Masaka, particularly at peak-harvest period, has been reported (District Agriculture Office reports). Since in this case the migrants usually provide casual labour. The impacts on the coffee estate firms are likely to be associated with the migrants as a likely source of infection to the permanent staff, as well as difficulty in predicting casual labour availability during peak crop activity.

HIV and AIDS Impacts on Social Institutions

HIV and AIDS is destroying the institutional fabric serving rural communities. Formal and informal institutions suffer when staff and members fall sick and die from HIV and AIDS- related illnesses. Starting with informal institutions, social capital has been described as one of the important assets of the poor because of its potential to act as a safety net in times of stress or crisis. Social capital refers to the strength of associational life, trust and norms of reciprocity (Haddad and Gillespie, 2001). Like other assets, social capital is damaged in a number of ways by HIV and AIDS. Some of these include: disruption of the process of social reproduction, diminished incentive to invest in norms of reciprocity, weakening of social institutions and networks as a result of HIV and AIDS-related mortality and increase in the number of orphans and widows that need support. As households contend with increasing expenditure (e.g. health care, funerals and fostering orphans), while earning less income, it becomes more and more difficult to mobilize local resources for communal or group-based initiatives. Groups may eventually disintegrate as members die, or can no longer contribute time or afford to pay their dues. Haddad and Gillespie (2001) further suggest that HIV and AIDS might lead to the generation of a type of social capital formation that is good for those intimately involved with the network, but which has negative externalities for non-members (i.e. social exclusion). Additionally, it is noted that social networks might be strengthened initially, thus stimulating collective action, but later become weakened as the impacts undermine the ability and incentive for collective action (see case below – Box 2).

Box 2

Saffina has been a widow for five years. Her husband died of AIDS in 2002 and she tested HIV positive in 2003. After her husband's death, Saffina joined a group of three other women in her villag (Lweza HIV/ AIDS Widows' Group) so that they could help each other by pooling labour for various agricultural activities. Saffina says that the group worked together very well till she started falling ill more frequently towards the end of 2004. In this is the period she developed TB and had no energy to work and reciprocate what her group members were doing for her. At first, say for the first four to five months of 2005, the friends continued to work in her gardens, either weeding or harvesting. After being on medication Saffina developed some strength but she was no longer as efficient as she used to be, or manage working the whole day in the hot sun. Gradually, the other group members stopped requesting her labour and when she would ask for help, each one would give an excuse. Apart from work, the women have even stopped visiting her. Her three boys aged 12, 9 and 7, are the only helping hands she now has. They do most of the housework, including digging in the mornings before going to school. When she is sick, the eldest boy stays at home to look after her. (Interview conducted by M. Karuhanga Beraho, in Masaka, 2005.)

In Uganda, different communities have responded differently. While some studies conducted in the western part of the country have reported development of informal institutions (where they were originally non-existent) to support people living with HIV and AIDS, the unsustainable strains on the extended family structure imposed by the orphan crisis cannot leave institutions in high-prevalence areas the same. Implications for agricultural development due to changes in social capital will vary, both within and between communities and farming systems, and are likely to depend on the level of organization, cohesion and strength of social networks before the epidemic, as well as on their role in ensuring food security. Nonetheless, the impact of HIV and AIDS on informal institutions among farmers has significant implications for agriculture, given that such institutions are likely to facilitate more effective targeting of farmers by both research and agricultural advisory agencies. Strengthening of social capital networks where they still exist may provide opportunities for increased agricultural production and therefore food security in the face of continued labour loss due to HIV and AIDS-related mortality.

Formal institutions such as agricultural extension and research agencies have also experienced the brunt of AIDS. First and foremost, these institutions have been progressively deprived of experienced people through HIV or AIDS-related deaths. The deaths of experienced professionals presents one of the key challenges in HIV and AIDS mitigation efforts. As Cohen, (2000 cited in Haddad and Gillespie 2001) pertinently says "How (do we) achieve sustainable development essential for an effective response to the epidemic under conditions where the epidemic is destructive of the capacities essential for the response?" Secondly, the productivity of the human resource left behind suffers due to repeated periods of illness that lead to recurrent absences from work, increased workload and reduced morale due to the loss of friends and colleagues. Thirdly, there is increased use of institutional resources to meet HIV or AIDS-related costs (for example, medical care, life insurance claims, burial costs, and time spent on burials) of the employees (Loevinsohn and Gilespie, 2003). The loss of institutional capacity and the expenses involved in coping with staff loss and death can undermine public and private sector service delivery and the sustainability of sector programs. These needs must be seen in light of the already existing staffing constraints in agricultural extension services and the high farmer: extension agent ratio experienced in the country. Even so, the consequences of this institutional breakdown may lead to a collective and individual inability to deal adequately with HIV and AIDS.

Gender dimensions of HIV/AIDS

HIV and AIDS impacts are not gender-neutral but are mediated by social-cultural landscapes (Gillespie *et al.*, 2001; UNDP, 2002). Women are often more susceptible to HIV infection and more vulnerable to AIDS impacts compared to men (UNAIDS, 2004). Following the death of a spouse, it has been found that widows suffer more than widowers. Inheritance customs, especially the traditional claiming of land and property by the deceased man's male family members, for example, has resulted in some orphans and widows being dispossessed of their parent's or spouse's inheritance, thus increasing the vulnerability of widows and orphans (Wakhweya *et al.*, 2002; UNDP, 2002). Wakhweya *et al.* (2002) further assert that disinheritance represents a major economic loss to orphans and their families, given that it not only reduces assets and income opportunities in the short term but also has implications for children's long-term economic security.

Table 9.1 *Summary of ways in which HIV and AIDS may affect agricultural growth*

Level of Impact	How does HIV/AIDS change the context of agricultural growth?	Leads to
Rural Household **A. *Human Capital***	*1. Labour changes* *Shortage of labour due to:* • Mortality • Morbidity – reduced productivity • Surviving adults take care of infirm *Shortage of hired labour due to:* • Mortality • Migration to cities • Lack of cash to pay for labour *2. Loss of agricultural skills and knowledge* • Premature mortality curtails period for intergenerational role modeling an knowledge transfer *3. Reduced quality of life*	• Less land being farmed • Underfarming of land in the absence of labour sharing and well-defined property rights • More child labour • Less labour-intensive crops grown • Emphasis on meeting food needs first and cash crops later • Greater emphasis on small livestock production • Decline for marketed output for crop processors • Natural resource mining (the future is heavily discounted) • Less appropriate farming practices in a more hostile environment More farmers who are inexperienced and need training, (e.g. youth); lack of role models • Increased burden of care and workload, particularly for women and girl children

Table 9.1 (contd.)   *Summary of ways in which HIV and AIDS may affect agricultural growth*

Level of Impact	How does HIV/AIDS change the context of agricultural growth?	Leads to
B. Financial Capital	*Income changes* • Fewer earners, increase in dependency ratio • Greater expenditure on medical, transport, special needs of the ill	• More off-farm income sources • Migration • Decrease in available cash spent on farm equipment and agricultural inputs
C. Physical Assets	*Changes in Assets* • Sale of assets • Loss of assets (e.g. land) by widow & orphans to spouse's relatives • Loss of unused land	*Changes in Assets* • Sale of assets • Loss of assets (e.g. land) by widow & orphans to spouse's relatives • Loss of unused land
Agricultural Estate Sector	1. *Labour Changes* 2. *Financial changes* - increased	• Reduced productivity • Loss of valuable skills and knowledge • Increased medical and funeral costs • Increased costs of recruitment and retraining of new staff

Table 9.1 (contd.)   *Summary of ways in which HIV and AIDS may affect agricultural growth*

Level of Impact	How does HIV/AIDS change the context of agricultural growth?	Leads to
Farming System	1. *Changes in cropping patterns* • Reduction in cultivated land and variety of crops grown 2. *Mortality & morbidity-related labour changes leading to* • Less appropriate farming practices; land becomes bushy 3. *Changes in livestock*	• Reduction in crop yields • Emphasis on less nutritious but less labour-intensive crops • Increase in prevalence and spread of plant and animal diseases • Reduced uptake of recommended agronomic practices • Shift to keeping of small livestock
Social Institutions	*Institutional and organizational changes* • Loss of institutional knowledge, high turn-over, low investment in staff development	• Weaker rural institutions(e.g. extension services, micro-finance institutions, NGOs) • Weaker social capital • Weakening of property rights for some • Weakening of asset base of women (especially land)

ADAPTED WITH MODIFICATIONS FROM: HADDAD AND GILLESPIE (2001)

This, coupled with stigmatization from in-laws, results in some widows becoming poorer, deeply distressed and with no motivation to live (UNDP, 2002). By contrast, widowers stay with their properties and sometimes acquire new partners. Additionally, because women have been the traditional care providers in Uganda, the burden of care for AIDS patients and AIDS orphans has also fallen on women (UNDP, 1995). A study conducted by Wakhweya *et al.* (2002) yielded similar findings. In this study 85 percent of single-parent orphan households were headed by females. The growing burden on women as they care for sick family members and orphans has serious implications for food production and consequently the food security status of affected households, as well as the community in general. There is also a growing number of households that have limited capacity to produce their own food, such as grandmother-headed households (with very many orphans) and those headed by children whose parents have died from AIDS. In all these households, the children are under pressure to engage in agricultural activities but lack or have limited knowledge about production. These factors, whether singly or in combination, exacerbate the already negative situation of food insecurity[20] in such households. It is also important to note that men and women tend to own assets and money separately, and tend to have separate income- earning activities (Jones, 1985; Doss, 1994; and Udry 1994). As Opiyo (2001) points out, it appears that the rise of prolonged sickness and care due to AIDS has reduced the time available for women to engage in income-earning activities, more so than for men.

Furthermore, HIV and AIDS exacerbate socioeconomic and cultural inequalities that define women's status in society (UNDP, 1992). HIV and AIDS for example worsens gender-based differences in access to land and other productive resources like labour, technology and credit (UNDP, 1995). Widow-headed households suffer from the direct loss of the spouse's farm labour and income, since men are the ones who usually work outside the home (UNDP, 1995 and Wakhweya *et al.*, 2002). In such circumstances, widows from wealthier households may be better off (in terms of resilience

---

[20] Available agricultural and population statistics data indicates that average per capita food production in 1999-2002 was 35 percent less than what it was in 1970-1972 (World Bank, 1993; UBOS, 2001). This also needs to be viewed in the context of Uganda's high annual population growth rate of 3.4 percent (UNDP, 2004).

to HIV and AIDS impacts) than to those from poor households. Furthermore, premature death of a spouse may also deprive the female of the necessary time to build up a set of extra family levers – such as access to community land, community groups, and/or microfinance groups – that can be used to exert power within the family (Gillespie *et al,* . 2001) but which are also vital for her livelihood and that of her children. Therefore, if property rights for a whole range of assets are not clearly and equitably defined or are not enforced, women are likely to be less able to shape their own destinies. It is important to mention that gender disparities and cultural practices also tend to render the girl child particularly exposed to exploitation and heavy responsibilities, especially in areas of housekeeping and agricultural production (Jayne *et al.,* 2005). Since female-headed households (widow and single women-headed) in Uganda have been identified as having the highest vulnerability to poverty ahead of male youth and households with large families, the superimposition of the devastating consequences of HIV and AIDS is likely to further compound their vulnerability (MAAIF, 2000).

The Challenge to Agricultural Extension Services

The foregoing sections contain clear evidence of the way the HIV epidemic has changed and continues to change the landscape of farming communities, which bear implications for agricultural extension services. In this section some of the major challenges likely to be (or already being) experienced are discussed.

Apart from the usual logistical problems of limited funding, the challenges that most agricultural extension agencies have faced in the past were mainly of a technical nature, relating to issues like dealing with serious pest and/or disease outbreaks, natural disasters (drought/floods) or intensive campaigns to promote certain technologies. However, the HIV epidemic, in combination with other prevailing technical, socio-economic and political factors have created a completely new set of complex challenges that extension agencies have to deal with. The new set of challenges could be categorized into the impact of the epidemic on the clientele of extension services and on the extension organization itself. A brief discussion of these impacts follows.

The HIV epidemic is changing the traditional composition of the clientele of extension services. In areas of high HIV prevalence, the category of productive men, women and youth in late adolescence to middle age, is one that has been most affected by the epidemic. One currently finds more women, children and elderly persons engaged in farming due to prolonged illness and/or death of their spouses, parents, guardians and other members of the family. Change of this magnitude in a clientele that was already resource-limited and mostly from illiterate rural households is bound to render the existing extension strategies and methods irrelevant unless they are adjusted in line with the new extension clientele and their needs (Qamar 2001, Qamar 2003). Furthermore, the changed structure is likely to have an effect on the level of appreciation of extension messages as well as on the efficiency of adoption of recommended practices, given that the very old and young may not be active seekers of extension information. The progressive depletion of household assets associated with the epidemic, increased expenditure on HIV and AIDS-related treatment and care, the use of an already strained household asset base to support many orphans in some households, as well as changes in farmers' ability to access and control productive assets and resources, have worked in different combinations to further make it more difficult for people to cope during other stressful situations like drought, flooding, or economic hardships that are common phenomena in this part of the world. Under such conditions of severe stress, short-term survival is a priority over any investments into agriculture or sustainable management of natural and other resources. More features of the changing landscape are associated with the increasing number of households headed by children, who have to work in agriculture but lack the necessary knowledge and skills to earn a living; increasing numbers of households headed by grandmothers who are too old/weak to work and/or have many orphans to look after; the growing HIV and AIDS-related care burden on women with consequences of reduced food production, and agricultural communities with a depleted local agricultural knowledge and experience. These are just a few examples of the changing context in agricultural communities but in no way provide the whole picture of the complex nature of changes that have taken place in people's livelihoods.

Regarding impacts at organizational level, key ones include the already-mentioned reduced capacity due to loss of skilled and experienced staff coupled with new institutional, technical and operational challenges. Presently, there is very limited tested extension approaches, methodologies and strategies to improve agricultural skills of inexperienced young farmers, including a large number of women and orphaned children who have suddenly become clientele of the services. Furthermore, the increased need to develop new technologies and equipment suitable for the new situation is difficult to address, given the weak linkages that have long existed between extension, research and other relevant agencies. Moreover, when one considers the content of extension messages, this remains strictly confined to agriculture. Perhaps the most challenging is that the extension staff who are supposed to drive the process, are themselves ill equipped to cope with the situation because of their lack of knowledge about AIDS. It is noteworthy that the impacts of the epidemic on other institutions also has consequences for extension agencies, particularly in terms of levels of funding received from government and the type and/or effectiveness of the collaborative activities engaged in. Despite the challenges, major efforts are needed by the extension agencies to reorganize and prepare themselves for the battle against the epidemic, because of their important role in rural development.

The Role of Agricultural Extension Services

Given their mandate to facilitate rural human resource development with the aim of increased food production through use of improved technology, agricultural extension organizations clearly have an important role to play in preventing and mitigating the effects of the epidemic. This is not to mean that they are expected to be medically involved in the fight against AIDS, but rather to emphasize their extremely important role in the management of the effects of HIV through provision of relevant and targeted farming knowledge and techniques. Below are examples of interventions where extension, training and advisory services can make (and are already making) an important contribution. The combination of interventions and how they are implemented will differ depending on the stage of the epidemic. The list is by no means exhaustive but merely aims to provide an overview of tried and tested methods that may be used to mitigate the epidemic's impacts.

Awareness Creation

The most meaningful role the extension services can play is strengthening the prevention of further spread of HIV infections by educating male and female farmers about the subject, and explaining to them the links between HIV and AIDS, food insecurity, malnutrition, agricultural production and rural poverty. Preparation of multimedia materials on HIV and AIDS could be combined with public awareness campaigns. FAO has produced a number of fact sheets on how to integrate HIV and AIDS into agriculture (http://www.fao.org/hivaids/) which could be adapted to various contexts. Local leaders in the communities could be specifically targeted for sensitisation, so that they can also be used in HIV and AIDS awareness campaigns. In Zambia and Uganda, the IP designed and piloted communication campaigns (see Box 3 below).

Livelihoods and food security support for HIV and AIDS affected households

The role of good nutrition in ensuring quality of life for people living with HIV and AIDS cannot be overemphasized. However, as in the section above, a mix of interventions may be required to assist households affected by HIV and AIDS. This mix may include supporting food production and introducing income generation, promoting labour-saving technologies, intensifying promotion of nutrition education programs, facilitating development of local food banks (although this has not been tried in Uganda), encouraging labour-pooling arrangements, where feasible, as well as reinforcing community-based mechanisms to preserve local agricultural knowledge. Attempts can be made to identify nutritious local plants that can be grown locally and used to improve the nutrition of affected individuals. Actions to promote better nutrition through on-farm processing and the utilization of indigenous crops should be encouraged. It is important to note that, sometimes, the mix of interventions may be combined with food aid for those households that have no capacity

Box 3

In Uganda, The Integrated Support to Sustainable Development and Food Security Program (IP) developed an HIV and AIDS resource guide for agricultural extension and community workers and building capacity of extension staff at the district level. The guide can be used during training and as a reference material while working with communities. It provides:

- a clear picture of the dynamics of HIV/AIDS in different agricultural production systems;
- practical skills for assessing and analyzing the nature of HIV/AIDS' impact on vulnerable groups;
- practical guidelines for working with communities to develop appropriate response strategies for HIV/ AIDS-affected households, for example, related to nutrition, labour-saving technologies, community mobilization;
- useful resources and ideas that can be adapted.

(IP & FAO, 2003)

to produce food. In this case, extension personnel would need to liaise with organizations that give food aid to obtain the necessary support for affected households. Most of the food aid that supported people living with HIV and AIDS has been imported into the country at very high costs, making such programs unsustainable. This presents an opportunity for promotion of local small-scale agroprocessing of foods to make high-nutrient products that would be suitable for people living with HIV and AIDS. In addition to providing opportunities for diversification in the community, these products would be produced at relatively lower costs and therefore programs providing food-relief aid would be able to support more people in need. It is important to note that the type of support given will depend on the household type and/or stocks of capital assets and knowledge base. Orphan and elderly-headed households, for example, may need more direct support and aid, while households fostering orphans may benefit from enhanced access to factors of production such as microfinance (at reasonable interest rates). It is therefore imperative that efforts are made to understand the specific needs of different household types rather than base support on a homogenous conceptualization of the sick or affected. This does not only improve good targeting of the most vulnerable, but also helps to address bias (including gender) with regard to the social groups that receive production and marketing knowledge or other opportunities.

Education, life skills and vocational training for AIDS orphans and vulnerable children

Many orphans and vulnerable children are unable to attend school. Even where incentive programs exist, few are educated beyond primary level since, in most cases, the conditions in the foster homes or factors like lack of parental care and guidance in child-headed households are not conducive. Extension can therefore participate in provision of life skills and vocational education especially that related to nutrition, food and agriculture. Such education and training could be achieved through both formal and informal channels. In Uganda a local NGO, Kitovu Mobile AIDS Homecare, Counseling and Orphan Program (MAHCOP) has been running a farm school for AIDS orphans and vulnerable children for the last three years. The program has also been modified to suit adults in HIV and AIDS-affected households that need extension advice (MAHCOP, 2004/5). The experience of MAHCOP can be used to design similar programs in other parts of the country. MAHCOP operates in the three districts of Masaka, Rakai and Sembabule. The Population Division (SDW) of FAO, together with the World Food Program (WFP) and other partners have also piloted Junior Farmer Field and Life Schools (JFFLS) in some African countries, designed specifically for vulnerable AIDS orphans and other vulnerable children, and Adult Farmer Field and Life Schools (AFFLS), designed to target poor households affected by HIV and AIDS, and especially adults in women-headed households (IP and FAO, 2003). While the success of all these programs is yet to be established, at least one certainty is that all of them are targeting extension information and support to unique extension clientele groups that were not targeted by traditional agricultural extension agencies. Additionally, formation of inter-country extension networks can be very useful in terms of experience sharing of how such programs are managed or could be strengthened to produce better impact.

Specific research on extension and HIV and AIDS

Although a number of studies on the relationship between HIV and AIDS and agriculture have been conducted (by FAO, for example), there is a need for extension agencies to engage in research focusing specifically on the relationship between HIV or AIDS and specific aspects of extension, in order to develop area-specific interventions. This needs to be done on a regular basis in order to keep track of trends in the epidemic's impact. Such studies can also feed into periodic reviews of existing agricultural policies and programs (particularly those on extension service delivery and food and nutrition security) to determine their relevance to the HIV and AIDS situation in Uganda. This calls for strengthening of inter-institutional collaboration, especially in the areas of agricultural research and extension. Additionally, the use of participatory research approaches in technology development and dissemination has been found to be useful and is recommended. The following could form some important areas of study (see Box 4):

Box 4

- Who is affected by HIV/AIDS, in what ways and why? In what ways are different groups responding? What are the implications for agricultural production? *What agricultural technologies and information would therefore be relevant?*

- What are the differential impacts of HIV/AIDS on men, women, children and the elderly (in terms of ability to work, health, nutrition)? *What are the implications for extension in terms of message design and content, as well as who gets targeted with what messages?*

- In what ways does HIV/AIDS influence or change access to and control over household resources by different household members? What are the implications for agricultural production and food security? *Given the differential impacts, what are the extension needs of different clientele groups? How does this affect the design and delivery of agricultural extension programs?*

- What is the effect of HIV/AIDS-related morbidity and mortality on household labour availability and household resource use? *What are the implications for relevance and adoption of new or existing agricultural technologies?*

- What local agricultural-related knowledge exists and who transmits it? *In what ways can extension agencies build upon such resources in the community?*
- How do poverty and food insecurity affect the spread of HIV/ AIDS within different communities? *How does this affect the effectiveness and efficiency of extension services?*
- In what ways does HIV/AIDS influence the relevance of (1) the technology development process (2) technology dissemination pathways and (3) the technology itself?-Which agricultural technologies have been abandoned and why, which technologies are affected households using and with what modifications? In what ways does HIV/AIDS influence a household's demand and uptake of new and/or existing agricultural technologies? *What are the implications with regard to extension service delivery strategies?*
- What is the potential of selected technologies for mitigating the effects of HIV/AIDS? *What is the role of extension?*
- What other environmental factors influence relevance of existing agricultural technologies? In what ways does HIV/AIDS reinforce these factors? *What are the implications with regard to extension service delivery strategy?*
- In what ways is the farming system changing in areas of high HIV prevalence? *What are the implications for agricultural technology development and dissemination?*
- *In what ways is the impact of the epidemic likely to undermine extension programs? What are the implications for extension service delivery?*

Conclusions

HIV and AIDS have had devastating impacts on the livelihoods of rural dwellers in Uganda, to the extent of threatening their very existence. HIV and AIDS-related impacts on the agricultural sector could be summarized as (1) labour shortage (2) asset depletion (3) knowledge loss and (4) a loss of formal and informal institutional support or capacity for collective action. Through these impacts HIV and AIDS produce different consequences for different individuals, communities and farming systems. In turn, individual or household responses influence the relevance and appropriateness of existing and new technologies as well as the process of technology development and dissemination. Although the vulnerability of households varies depending on different combinations of causative factors, a wide asset base and opportunities to diversify activities are likely to assist in protecting a household against external shocks, such as HIV and AIDS and adverse weather conditions.

In spite of prevailing challenges associated with changing extension clientele and institution context, the role of extension and other advisory agencies in educating the farming population about the disease, as well as in developing new strategies, methodologies, materials, technology and equipment to address new extension needs can be significant. As HIV and AIDS-related labour loss continues, there is a need to develop new cultivation technologies and varieties that are drought-resistant and nutritious but do not rely on much labour. The challenge at hand, though, is to develop agricultural technologies that adapt to the reality of HIV and AIDS affected environments while maintaining productivity, as well as developing dissemination pathways that are relevant. It would also be desirable if such technologies not only mitigated current impacts of HIV and AIDS but also have the potential to reduce susceptibility to future infection and vulnerability to various types of impacts. A number of interventions related to how HIV and AIDS could be integrated in agricultural activities have been discussed. The role of farmer field schools to facilitate the transfer of community-specific and organization-specific knowledge within and across generations, thereby mitigating knowledge-loss-related impacts was also highlighted and its practicability and/or up-scaling can be explored in different communities and under different circumstances.

All the above will obviously require additional human, financial and physical resources, review of existing curricula at agricultural training institutions, appropriate training of staff, development of integrated extension messages and appropriate strategies, partnerships with relevant institutions, and, above all, appreciation of the new situation among extension agents and the establishment of fresh working relationships with clientele. It is therefore imperative that national extension systems receive the necessary financial and technical assistance, first in view of badly needed HIV/AIDS education for the rural masses and secondly the increasing danger of food insecurity and poverty due to reduced agricultural production. Furthermore, and as

Qamar (2001) observes, the current trend to reform extension organizations through privatization of extension services in Africa may make sense under normal conditions but under current circumstances, its justification needs to be re-examined earnestly.

The Government of Uganda recognises the seriousness of the HIV and AIDS issue and has demonstrated its commitment through the establishment of a multi-sectoral approach to HIV and AIDS control. The National Strategic Framework (NSF) guides the implementation of all policies on HIV and AIDS and its mainstreaming into the development of sector policies. However, beyond policy and strategies there is a need for government to establish mechanisms to deal with existing government failures that have continued to increase rural people's vulnerability to HIV and AIDS, because that is when the effectiveness of efforts by agricultural extension agencies can be realized.

References

Asingwire, N. and Kyomuhendo S., 2003. An Analysis of Policy Gaps. Report compiled for Uganda AIDS Commission.

Barnett T. and Blaikie, P. (1992): AIDS in Africa: Its present and future impact. London: Bilhaven Press.

Barnett, T., Tumushabe, J., Bantebya, G., Sebuliba, R. Ngasongwa, R. Kapinga, D., Ndelike, M., Drinkwater, M., Mitti, G., and Haslwimmer, M. (1995): The Social and Economic Impact of HIV/AIDS on farming systems and Livelihoods in Rural africa: Some experiences and Lessons from Uganda, Tanzania and Zambia, *Journal of International Development* 7, p.p. 163 – 176.

Barnett T. and Whiteside, A. (2002). Poverty and HIV/AIDS: Impact, Coping and Mitigation Policy. In G.A. Cornia (ed.) AIDS, Public Policy and Child Welbeing, Chapter 11. Florence:UNICEF. http://www.unicef-icdc.org/research/ESP/aids/chapter11.pdf.

Bebbington, A. (1999): Capitals and capabilities: a framework for analyzing peasant viability, rural livelihoods and poverty. *World Development*. 27 (12), p.p. 2021-2044.

Doss, C. (1994): Models of Intrahousehold Resource Allocation: Assumptions and Empirical Tests. Staff Paper P94-18. St. Paul: Department of Agricultural and Applied Economics, University of Minnesota.

Du Guerny, J. (2002): Agriculture and HIV/AIDS. New York: United Nations Development Program (UNDP).

FAO (Food and Agricultural Organization of the United Nations) (2003): The impact of HIV/AIDS on the agricultural sector and rural livelihoods in Uganda. Baseline report, August 2003. Kampala, Uganda: FAO.

FAO (Food and Agricultural Organization of the United Nations) (2001): The impact of HIV/AIDS on food security. Rome: FAO.

FAO (Food and Agricultural Organization of the United Nations) (1992): The effects of HIV/AIDS on farming systems in eastern Africa. Rome: FAO.

FAO (Food and Agricultural Organization of the United Nations) (1995): The effects of HIV/AIDS on farming systems in eastern Africa. Rome: FAO.

Gillespie S., Haddad L. and Jackson R. (2001): HIV/AIDS, Food and Nutrition Security: Impacts and Actions. Paper prepared for the 28th Session of the ACC/SCN Symposium on Nutrition and HIV/AIDS.

Global fund (2005): "Global Fund Suspends Grants to Uganda", Global Fund Press Release, 24 August 2005 (Down loaded February, 27, 2007)

Haddad L. and Gillespie S., (2001): Effective Food and Nutrition Policy Responses to HIV/AIDS: What we know and what we need to know. *Journal of International Development,* 13, p.p. 487-511.

Hurst, P., Termine, P. and Kare, M. (2005): Agricultural Workers and their Contribution to Sustainable Agriculture and Rural Development. Rome: FAO-ILO-IUF, 2005.

IP & FAO, 2003. HIV/AIDS and agriculture: impacts and responses Case studies from Namibia, Uganda and Zambia. Rome: FAO.

Jayne T.S., Villarreal, M., Pingali, P. and Hemrich, G. (2005); HIV/AIDS and the Agricultural Sector: Implications for Policy in Eastern and Southern Africa. *Electronic Journal of Agricultural and Development Economics* , Vol. 2, (2), p.p. 158-181.

Jones, C. (1986): Intra-Household Bargaining in Response to the Introduction of New Crops: A Case Study from North Cameroon. In J. Lewinger Moock, (ed). Understanding Africa's Rural Households and Farming Systems. Boulder: Westview.

Karuhanga, Kwesiga (1998) (n.d.), Loevinsohn M. and Gillespie, S. (2003): HIV/AIDS, Food security and rural Livelihoods: Understanding and responding. RENEWAL Working Paper No. 2. May 3003.

MAHCOP, (2005): Kitovu Mobile AIDS Homecare, Counseling and Orphan Program, Project Report. Kitovu, Uganda: MAHCOP.

Marenya, P., Oluoch-Kosura, W., Place, F. and, Barrett, C. (2003): Education, Non-farm Income, and Farm Investment in Land-scare Western Kenya, BASIS Brief 14, University of Wisconsin, Madison: Department of Agricultural Economics.

MAAIF (Ministry of Agriculture Animal Industry and Fisheries) (2000): National Agricultural Advisory Services (NAADS) Program. Master document of the NAADS Task force and joint Donor group. Entebbe: MAAIF.

MFPED (Ministry of Finance, Planning and Economic Development) (2003): Uganda Poverty Status Report, 2003. Kampala, Uganda: MFPED.

MFPED (Ministry of Finance, Planning and Economic Development) (2004): Poverty Eradication Action Plan 2004/5 – 2007/8. Kampala, Uganda; MFPED.

Ministry of Health (MOH) (Uganda) and ORC Macro (2006): *Uganda HIV/AIDS Sero-behavioural Survey 2004-2005*. Calverton, Maryland : Ministry of Health and ORC Macro.

Mongi, R. (2002): Assessment of the Impact of HIV/AIDS Epidemic on the Coffee-Banana Farming system in Aremeru District, Tanzania. Deventer: Larenstein University of Professional Education.

NEMA (National Environment Management Authority), (1998): State of the Environment Report for Uganda 1998. Kampala: NEMA.

NEMA (National Environment Management Authority) (2001): State of the Environment Report for Uganda 2000/2001. Kampala: NEMA.

Niehof, A. and Price, L. (2001): Rural Livelihood Systems: A Conceptual Framework. WU-UPWARD Series on Rural Livelihoods, No. 1.

Qamar, M.K. (2001): The HIV/AIDS epidemic: An unusual challenge to agricultural extension services in sub-Saharan Africa. *Journal of Agricultural Education and Extension* 8, (1), 2001, pp 1-11.

Qamar, M.K. (2003): Facing the challenges of an HIV/AIDS epidemic: Agricultural extension services in Sub-saharan Africa. Rome: FAO.

Opiyo, P. (2001): HIV/AIDS, Gender and Livelihood in Siaya District, Kenya: An Analysis of AIDS Impact on Rural Households. MS. Thesis. Wageningen, Netherlands: Wageningen University.

Reardon, T., Crawford E., and Kelly, V. (1995): Promoting Farm Investment for Sustainable Intensification of Agriculture, International Development Paper 18, East Lansing: Department of Agricultural Economics, Michigan State University.

Rugalema, G. (1999): Adult mortality as entitlement failure: AIDS and the crisis of rural livelihoods in a Tanzanian Village. Ph.D. Thesis. Den Haag: Institute of Social Studies. Semana 1999.

Tanzarn, N. and Bishop-Sambrook, C. (2003): The dynamics of HIV/AIDS in small-scale fishing communities in Uganda. Draft report of a study. Rome: FAO.

The Lancet, 366(9500), 26 November - 2 December 2006. "Uganda is learning from its Global Fund grant suspension".

Topouzis, D. (2000): Measuring the Impact of HIV/AIDS on the Agricultural Sector in Africa. Geneva: UNAIDS Best Practice Collection.

Udry, C., (1994): Gender, Agricultural Productivity and the Theory of the Household. Journal of Political Economy, 104(5), pp. 1010–1046.

Uganda AIDS commission, 1997. The HIV/AIDS Epidemic: Prevalence and Impact. Kampala: Uganda Aids Commission.

Uganda AIDS commission, 2003a. The HIV/AIDS Epidemic: Prevalence and Impact. Kampala: Uganda Aids Commission.

UAC (Uganda AIDS Commission) (2003b): The National Strategic Framework for HIV/AIDS, Activities in Uganda: 2000/1 – 2005/6. Mid-Term Review Report .

UAC (Uganda AIDS Commission) (2006): The Uganda HIV/AIDS Status Report. July 2004 – December, 2005. Kampala: Uganda Aids Commission.

UBOS (Uganda Bureau of Statistics) (2003a): National Household Survey 2003, Report on the labour force survey. Kampala, Uganda: UBOS.

UBOS (Uganda Bureau of Statistics) (2003b): 2002 National Census Report.

UNAIDS - XV International AIDS Conference Bangkok UNAIDS 2004 Report on the global AIDS epidemic. www.unaids.org/bangkok2004/report.html. (Accessed December, 2006)

UNAIDS, 2006 Report on the global AIDS epidemic.

UNDP (United Nations Development Program) (1995): Quoted in Human Development Report 1997. New York: Oxford University Press.

UNDP (United Nations Development Program) (2002): Uganda Human Development Report 2002. The Challenge of HIV/AIDS: Maintaining the Momentum of Success. Kampala: UNDP.

Yamano, T. and Jayne, T.S. (2004): Measuring the Impacts of Working Age Adult Mortality among Small-Scale Farm Households in Kenya. *World Development,* 32(1), January.

Wakhweya, A., Kateregga, C., Konde-Lule, J., Mukyala, R., Sabin, L., Williams, M. and Heggenhougen, H.K. (2002): Situation Analysis of Orphans in Uganda. Orphans and their households: caring for their future today. Final Report. November. Kampala: Ministry of Gender, Labour and Social Development.

World Bank. 1993. Uganda: Agriculture. Washington, D.C: World Bank.

http://www.cdc.gov/nchstp/od/gap/countries/uganda.htm (down loaded February 27, 2007)

www.fao.org/hivaids (Accessed December, 2006)

www.unaids.org/bangkok2006/report.html (Accessed December, 2006)

www.ingramcontent.com/pod-product-compliance
Lightning Source LLC
Chambersburg PA
CBHW021049210326
41598CB00016B/1146